Digital Defense

Joseph N. Pelton • Indu B. Singh

# Digital Defense

*A Cybersecurity Primer*

 Springer

Copernicus Books is a brand of Springer

Joseph N. Pelton
Executive Board
International Association for the
  Advancement of Space Safety
Space and Advanced Communications
  Research Institute
Arlington, VA, USA

Indu B. Singh
LATA's Global Institute for Security Training
Los Alamos Technical Associates
McLean, VA, USA

ISBN 978-3-319-19952-8      ISBN 978-3-319-19953-5   (eBook)
DOI 10.1007/978-3-319-19953-5

Library of Congress Control Number: 2015947778

Springer Cham Heidelberg New York Dordrecht London

Cover image used courtesy of Flickr user Chris Halderman through a Creative Commons license

Printed on acid-free paper

Springer International Publishing AG Switzerland is part of Springer Science+Business Media (www.springer.com)

*This book is dedicated to the hardworking cyber security community that seeks to develop antivirus software, firewalls, and protective systems to defend against hackers and cybercriminals that would invade your digital world. We hope that this book can help to save would-be targets of cybercriminals and that the advice in this book will help to stem attacks by those that seek to use the Internet for ill-gotten gain and other nefarious purposes.*

President Obama's Official Statement of February 13, 2015, on Cybersecurity and its strategic importance to the United States:

America's economic prosperity, national security, and our individual liberties depend on our commitment to securing cyberspace and maintaining an open, interoperable, secure, and reliable Internet. Our critical infrastructure continues to be at risk from threats in cyberspace, and our economy is harmed by the theft of our intellectual property. Although the threats are serious and they constantly evolve, I believe that if we address them effectively, we can ensure that the Internet remains an engine for economic growth and a platform for the free exchange of ideas.

Statement given on the occasion of the U.S. Cyber Security Summit, February 13, 2015

https://www.whitehouse.gov/.../president-obama-speaks-white-house-summit_on-cyberSecurity.htm

# Preface

Cyber-attacks are increasing exponentially in the United States and around the world. Attacks in the United States are now averaging over 550,000 per week, or over 25,000,000 per year. Annual attacks on official U.S. government Internet sites have doubled from 31,000 in 2012 to over 60,000 in 2014. The increase of cyber-attacks is like an epidemic, and the threats to those that are linked to the Net via a desktop computer, a mobile phone, or a wireless local area network (LAN) in an office or router in their homes are of real concern.

In this short book there is straightforward and practical advice about how to defend yourself and your family against these often unprincipled and indeed criminal attacks. If you have an elderly mother or father or grandparent who uses the Internet you might buy this book for them. Or perhaps more likely buy it and go over the most relevant parts with them and arrange with them to purchase at least basic firewall, antivirus, and identity theft protection and set them wise to key Internet scams to look out for.

We hope this book provides lots of useful advice and good counsel. But we believe it can be most useful in helping you to defend your children, your immediate family, and especially elderly family members against the increasingly sneaky trick of cyber criminals. It provides assessments of various cybersecurity offerings and tips on strategies about how to go about obtaining professional assistance from competent computer security firms. The small and usually reasonable annual fees these companies charge can provide you with cyber protection that amounts to far less than the losses you might incur if you do not take these precautions.

Key elements that you will find in this book include:

• A clear and understandable presentation (i.e., no "techno-speak") of the various types of cyber threats that can now come to you via your Internet connections. These include, but are not limited to, viruses that can infect and disable your computer, malware that can allow your computer to send spam (unwanted e-mails out to thousands of others) for many nefarious purposes, and other types of computer trickery you should look out for. These computer tricks by "black hat hackers" (sometimes called "crackers") keep growing. Computer criminal are getting sneakier. Such antics include what are called "phishing" and "pharming." These seemingly legitimate messages are actually from computer criminals and might lead you to give away important passwords that could result in your financial accounts being drained of money, or perhaps worse.

We also provide advice about coping with "ransomware," data bombs, and Trojans—all of which are dangers to watch out for in today's world. We are moving into a new world sometimes called the "cyber-crime bazaar." These are dark networks where a wide range of cybercriminal activities are conducted. Here you can buy stolen credit card numbers, buy kits to steal information off of wireless instant pay cards, or even find ways to disable alarms and other protective electronic systems. Keyless entry systems are just one of the new frontiers for cyber criminals.

• Practical advice about how to protect yourself and your family from cyber criminals that are variously called "black hat hackers," "crackers," or simply "hackers." We emphasize that you might find this book useful to defend not only yourself but also your spouse, your children, and perhaps most of all elderly parents who are hip enough to go online, write and read e-mail, use a smart phone, and even send out texts and Tweets but may not be the most adept at defending themselves against cyber scammers.

- In today's world, where baby monitors and home security systems can be hacked, smart refrigerators and washing machines can send out spam, and cyber thieves with scanners can roam a neighborhood seeking out unprotected wireless routers so they can hack into bank savings and stock broker accounts, you need to be equipped to know how to protect you and your loved ones against those who would use the Internet and other electronic systems to extort money, empty accounts, capture key credit card or social security information, or steal identities.
- Information about the latest professional services that address cyber-bullying. This abuse of the Internet has become almost an epidemic in the last few years, with dire results such as public humiliation and even teen suicides. There are professional cyber-services one can obtain not only to protect yourself against cybercriminals but also to deal more effectively with cyber-bullying. These services allow those attacked by cyber-bullies to report such attacks and bring those that abuse the Internet to justice.
- An explanation of what is "identity theft" and why this is perhaps one of the worst things that can happen from a cyber-attack. This is because it could not only expose you to substantial financial loss and a very long hassle to correct the problem, but you could end up being charged with crimes that you did not commit because some criminal has assumed your identity.
- There is an up-to-date listing of various computer security services that can offer protective services against cyber criminals. These include providers of such services as "antiviruses," firewalls, password protection, and insurance against identity theft. Although not foolproof these services can go a long way to protect you against cyber-attackers. There is also information on more professional ways to track down those that would seek access to your wireless computer routers and wireless LANs without authorization.
- We provide you a rundown on why you need to be careful when you access the Net via "smartphones" and precautions that you should take when you sign up for automatic online "pay and go" or "tap and go" services such as "Apple Pay," "Blink," American Express "Express," etc.
- There are also some more detailed chapters about vital infrastructure. These chapters discuss other things that some users need to be concerned about. This is because the security of these "hidden digital systems and vital infrastructure" are now key to our daily lives. We depend on digital infrastructure that has widespread impact on lives. Thus we provide up-to-date information for those people who are dependent on satellite services to attain broadband access, GPS navigation, and other space-based services.

There is also a discussion of security issues involving the use of "the Cloud" because the government and more and more companies use the Cloud to store our vital information, process our tax returns, and keep track of our bank accounts.

• Finally we discuss briefly the specialized computer communications networks that control electrical grids, traffic signals, pipelines, water supplies, sewage treatment, and other urban infrastructure. These digital systems are known by the catchy name of "Supervisory Control and Data Acquisition (SCADA)" networks. Many might think that they do not need to know about such things, but it turns out that vital services you depend on from banks, local governments, the federal government, power companies, and more are potentially at risk with this type of cyber hazard. This means that you, too, are at risk. If you find these two chapters on vital infrastructure (i.e., satellites, "the Cloud" and "SCADA systems" in particular) turn out to be more detailed and involved than suit your taste, you can skip over them and proceed to the chapters about the future and the ten essential rules to follow.

Unfortunately there are others out there whose ambitions go beyond stealing money electronically or pulling pranks on people via the Internet. These are techno-terrorists that are seeking ways to use the Internet, information networks, remote and automated control systems, satellite links, or other electronic means to invade key governmental or military data banks. These techno-terrorists are conspiring to launch cyber-attacks against entire communities or even nations. Efforts to stop these sophisticated cyber-attackers, located in countries such as North Korea or within such terrorist organizations as ISIS, will dominate defense efforts more and more in future years.

One of the many problems with cybersecurity is that there are hundreds of terms that computer geeks use in this fairly technical field. To assist you there is a fairly detailed list of terms provided in the glossary to help explain the meaning of acronyms and to explain terms such as "whale phishing" and "near field communications" that are used in the new "tap-and-pay" systems, etc. Our goal, however, has been to use as few of these "techno-speak" terms as possible.

At the end of the book are some appendices that spell out the vital cybersecurity programs that are now being implemented in the United States and aboard for those that would like to know what their governments are doing to protect them against both cyber criminals and, even worse, techno-terrorists that attempt to carry out devastating cyber-attacks.

We have tried to be as comprehensive as possible in addressing the concerns that an individual or small business might have regarding cybersecurity and attacks that cyber-thugs might launch against you or your family. We have tried to explain basic cyber-risks and protective strategies without becoming enmeshed in techno-speak and gobbledgook terms that get in the way of a clear understanding of what the problems and solutions are. We hope you enjoy *Digital Defense*, which is designed to become your basic guide to cybersecurity. This is a book devoted to protecting you, your family, and especially seniors against those that abuse the Internet and digital technology.

Washington, DC, USA                                               Joseph N. Pelton
October 2015                                                          Indu B. Singh

# Acknowledgements

This is to provide acknowledgement for the detailed contributions and corrections of Alexander Pelton, our editor, who substantially helped to perfect this book. We wish to acknowledge the detailed contribution and corrections to the manuscript provided by Alexander Pelton. As our editor he helped to perfect this book. Also we wish to thank Peter Marshall for his review, comments, and suggestions. As always is the case, all errors are the responsibility of the authors.

# Contents

# 1

# What Is at Stake?
# What Should You Do?
# Why Should You Care?

If you and your family are accosted by a mugger on the street, then your money or your lives could well be at stake. If you live in Ukraine and Russian-backed invaders take over your town, then your livelihood or your home are likely to be in immediate danger. When a foreign power invades your country, you clearly know that you are most definitely in peril and that you had better fight back to defend your rights and that of your community.

Cyber-crime, however, can be so subtle and hidden, people can ignore the threat until it is too late. Yet today about every 3 s a person is hit by some form of cyber-attack out of the blue. It is estimated that a cyber-attack on the electric grid on which we all depend comes about once a minute. Precautions need to be taken up front to combat cyber-fraud, cyber-attack and most definitely cyber-terrorism. Locking the "cyber-barn door" after a "black hat" hacker has struck is way too late.

And sometimes the threat to you and your family might not be so subtle after all. Here are some frightening case studies in cyber-stalking, cyber-crime, and worse.

## Houston, Texas, and the "Hacked Baby Cam" in the Nursery

DATELINE JANUARY 28, 2015: A nanny named Ashley Standly was looking after 1-year-old Samantha in Houston, Texas, when she had a terrifying moment. It started when she heard a noise near the baby and walked over to investigate. Ashley could not believe her ears. A strange man's voice came

© Springer International Publishing Switzerland 2015
J.N. Pelton, I.B. Singh, *Digital Defense*, DOI 10.1007/978-3-319-19953-5_1

**Fig. 1.1** Security cameras that are connected to a home Wi-Fi Internet systems can be hacked

though on the baby's monitor. This particular baby cam had a microphone and a high resolution camera linked to the household's Wi-Fi Internet system. A strange man from nowhere could be heard calling the little girl "cute." The Wi-Fi wireless network had been used by a cyber-lurker to invade the privacy of this Houston household. What had been installed as a safety feature, but without security password protection, had become a portal for a digital voyeur to intrude into the nursery. This feat could unfortunately be achieved via a widely-available smartphone or wireless network app [1] (Fig. 1.1).

## The Sum of 22,000 lb Transferred from Elderly Mother's Bank Account in London, England

DATELINE FEBRUARY 28, 2015: A 31-year-old Nigerian hacker living in Southsea, England, used special software to access the e-mail account of Ilaria Purini, who lives in London. This is how he learned her passwords and the personal details of her life and her close relationship with her mother, who lives in Italy. In this case it turned out that Ilaria Purini worked for a museum and purchased art for her mother and sent it to her in Italy. Posing as Ilaria Purini in an e-mail the hacker sent instructions to a banker in Italy, who manages the bank account for Rosanna Rose, who is Ilaria's mother. He daringly ordered the transfer of money for the purchase of a work of art using earlier such orders as the model for the bank instruction. The first transaction was for 7784.59 lb, and when this transfer was successfully made to the account of the Nigerian hacker—rather than to Ilaria Purini—he doubled down and struck again. The second time he ordered the transfer of 14,215 lb. Eventually they were able to track him down, and the 22,000 lb were eventually regained. In this case it was not only getting the passwords and bank account numbers and bank manager's e-mail address, but the knowledge that art purchases were being made with orders to the bank manager for currency transfers [2].

## Couple Sentenced for False Tax Refund Conspiracy

Dateline April 24, 2014: In Charlotte, North Carolina, Senita Birt Dill and Ronald Jeremy Knowles were sentenced to multi-year prison terms and also ordered to pay $3,978,211 in restitution to the IRS. Dill and Knowles pleaded guilty to Internet access fraud and conspiracy charges. Dill also pled guilty to aggravated identity theft. Dill and Knowles used fraudulently obtained personal identification information (including names, dates of birth, Social Security numbers, and other personal information obtained via electronic and other means) to file false tax returns and claim falsified tax refunds. Dill and Knowles used neighbors' addresses to fill out the fraudulent tax returns and checked the homes' mailboxes frequently to retrieve the fraudulent refund checks upon delivery. The defendants also used addresses in Greenville and Greer, S.C., which belonged to Knowles' businesses. Dill and Knowles managed to file over 1000 false tax returns using the fraudulently obtained personal identification information before they were caught. These actions exposed over a 1000 people to potential charges of defrauding the U. S. government [3].

## Teens and Cyber Identity Theft

In some cases it turns out that teens are a perfect target for identity theft. Teens have a Social Security number and usually will have a perfect credit record, since they do not have debt; if there is a credit card it is often paid by their parents. It is only when teens go off to college or apply for their first credit card that they may find out that their identity has been stolen and used for some cyber-crime or fraudulent return [4].

## Kaitlin Jackson's Campaign Against Cyberbullying Trolls

Kaitlin Jackson of West Wales joined a support group on Facebook called Angel Mums after suffering the trauma of a miscarriage. But soon after joining, the group was overrun with tormenters called cybernet "trolls" who bombarded the women with horrifying messages of abuse, posting pictures of aborted fetuses and making jokes about dead children. Even worse she checked the Facebook and Twitter accounts used by her children and was appalled to find several examples of them also being targeted by these cyberbullies as well. "The comments sent to my were horrendous."

Kaitlin Jackson decided to fight back. Today she spends up to 8 h a day rooting out the identities of those responsible for placing contemptible messages on tribute pages and websites used by charity groups. If she finds out a troll's name, a Facebook profile page, or a website address she e-mails the details straight to the people responsible for policing the sites where the abuse appears. She also e-mails the trolls directly, warning them she is handing their messages over to police. Kaitlin Jackson is now a leading member of "Stop Cyberbullying and Trolls." Kaitlin's tactics involve bombarding the bullies with messages and asking them to explain their actions. Among the first trolls she tackled were those who created Facebook pages celebrating the murder of 5-year-old April Jones at Machynlleth, Wales, in October 2012. She now reports 40 people or sites to the police hate crime unit a day. The problem with criminal prosecution in many cases, however, is that messages posted on foreign sights are not currently subject to charges, especially if it cannot be proven that the cyberbullying attack was made by someone of local origin because of their totally anonymous postings. An international convention on cyberbullying would assist to combat trolls and their abusive behavior around the world.

Cyber-attacks that involve content (cyberbullying, racial, religious, or sexual orientation bigotry, pornographic sites, sexting and other such abuses of the Internet) are a difficult area from a legal viewpoint. Some of these parallel abusive activities such as racial, religious, and sexual orientation attacks, political extremism, inciting to violence, and instruction in terrorist attacks can be prosecuted as "hate crimes," but only through the court and police systems. There are also other forms of sexual deviation and pornography that are subject to yet other laws.

Some strategies to deal with these abuses of the Internet and social media that we hope will be found helpful are addressed in the following chapters. The most thorny problem associated with these types of attacks is that they don't involve manipulation of protocols and web address. When one is dealing with viruses, malware, Trojans, ransomware, identity theft, phishing and pharming, all of these things can be addressed through protective software and services. Your own computer or cell phone can be set to work to protect you and your family. However, when cyber-attacks involve what might be called "negative content" you enter the sphere of formal legal processes. Some hurtful activities carried out on the Internet are protected by freedom of speech. These abuses can only be resolved through a court of law and legal proceedings. Neither individuals nor computer software can sort out what is a criminal act. No software that you can install on your computer or cell phone can thus protect you from hate crimes or cyberbullying. We would note, however, it is on pornographic and hate crime sites that viruses or worms are often found to lurk [5].

## Hacking into Aircraft Communications and IT Wireless Communications Networks

In mid-April 2015 computer security expert Chris Roberts was banned from flying on United Airlines after he had used his laptop to hack into the aircraft's internal communications network and downloaded key flight-control and ranging information into a wireless hard drive memory system. United Airlines' spokesman Rahsaan Johnson said in explanation: "Given Mr. Robert's claims regarding manipulation of the aircraft system, we've decided it's in the best interest of our customers and crew members that he not be allowed to fly United." In this case, fortunately, Chris Roberts was looking for aircraft vulnerabilities that might exist due to online hackers in order to protect *against* such digital incursions. The FAA has urged that Boeing, in designing its 787 Dreamliner, design its communications network so that the flight control network can be completely segregated from customer networking systems to prevent such hacker attacks [6].

The above actual stories about cyber-attacks around the world unfortunately are just a few examples of the types of attacks via the Internet that might beset you and your family. Some of these attacks are easier to defend against than others. In Chap. 2 and following we try to help you defend your family against the dark side of the Internet.

Cybersecurity, cyber-crime, and cyber-terrorism are harder to understand and defend against than physical crime, but the threat to you, your family, your community and your country may be just as real and devastating. To understand the problems of cyber-security, we have tried to avoid jargon, but some technical terms are unavoidable. We have therefore provided a glossary and acronym guide at the end of the book. This should help to define terms that may be new to you.

Let's start with basics. There are three different levels of cyber-attacks. It is useful to understand these three major categories that experts have now labeled as the different ascending levels of cyber-attacks [7].

## Level One Attacks

Level One is a personal attack on yourself and/or your family. This is typically a cyber-criminal seeking personal gain at your expense. These types of attacks might include the following:

* Identity theft, fraud, and extortion
* Pharming, phishing, spamming, spoofing (These concepts plus malware, spyware, Trojans and viruses mentioned just below we will explain in just a moment.)
* Installation of malware, spyware, Trojan horses (or simply Trojans) and viruses
* Stealing of laptops or mobile devices to obtain passwords
* Denial-of-service or distributed denial-of-service attacks
* Breach of access
* Secret access to wireless LANs and Wi-Fi nets
* Password sniffing
* System infiltration
* Website defacement
* Intellectual property (IP) theft or unauthorized access
* Cyberbullying by trolls against yourself or members of your family
* Online "hate crimes" and other abuses through social media
* Internet pornography, sexting, and deviant behavior depicted online

## Level Two Attacks

Level Two is an attack on companies and community institutions and infrastructure. This could be everything from an attack by a disgruntled customer or employee against a company or bank to a major assault on a company's records, financial resources, or core data, to even a cyber-terrorist attack against a corporation providing vital infrastructure such as electrical power or telecommunications, or something like the North Korean attack on Sony Pictures.

## Level Three Attacks

Level Three is an attack on a national government that can be the moral or actual equivalent to an act of war. North Korea's attack on Sony was somewhere between a Level 2 and 3.

This book is primarily about protecting you and your family against level one attacks. The advice is primarily to protect your bank accounts, your brokerage accounts, your Social Security benefits, your payroll deposits, and your financial assets. But note this, and note it well. Cyber-terrorists and cyber-criminals could attack your community or nation. There is a danger of attacks on electrical power systems, oil pipe lines, transit systems, or water and sewage systems. Such cyber-attacks can create horrific results. Stealth cyber-attacks can be mounted via cleverly concealed and elaborately obscured telecom and IT networks. You may never have heard of something called a supervisory control and data acquisition (SCADA) network. But an attack on these networks is disturbingly easy and has devastating consequences. The wrong instruction through these networks with criminal or terrorist intent can cripple your community and put you and your loved ones at mortal risk.

This means that you need to act *now* to protect your assets against cyber-attacks. There are many easy and low cost steps that can be taken to protect you and your loved ones from cyber-attack. This is one of the key takeaways from this short do-it-yourself book.

There is another important message. You should support actions taken by your local, state and national government to protect your community against deadly cyber-attacks. If your bank account is safe, but your water is poisoned by sewage or your nearby nuclear power plant explodes, you won't be around to spend that money. Neither companies nor governments have yet done enough to protect against Level Two and Level Three attacks. This is because the financial costs can be high, the technological challenges daunting, and there is always the hope—the increasingly forlorn hope—that any company, agency or even country will be spared the most egregious or technically sophisticated attack.

# Level One Threats to Be Taken Seriously

Let's review risks that matter to you the most—Level One cyber-attacks.

Level One personal attacks that could darken your door or lighten your bank account are unfortunately numerous and diverse. Such Level One attacks can also result in stolen identity, ruined credit rating, and damaged reputation in personal and professional life. The amount of time, money, and legal resources required to try to recover from a successful cyber-attack can be very substantial. Such attacks thus must be taken quite seriously, and preventive actions are now essential. The ingenuity of cyber-crooks seems to know no bounds. Just a few examples of cyber-tricksters at work may help to understand why you need to be on your guard.

## *Handle with Care!*

Most people today start their day by going to their computer or smart phone to read their latest e-mails. Many of these messages contain an offer of a gift card or a free trip or another attractive come-on. Ninety-nine out of a hundred of such messages and their attachments are not dangerous. A few may even be legitimate offers or could be sharing information or coupons or financial benefits that you really want to have—and not merely be marketing spam. Most of these e-mails are in pursuit of what is called a "cookie." A cookie is your actual e-mail address in the language that the Internet routing system uses. Once your cookie has been captured this information can be passed on to marketing organizations. A smaller percentage are trying to invite you into what might be called a "cyber-lair" for more nefarious reasons. These "black hat" hackers are out to do you harm. When you open their message it will likely cost you something. This might be as minor as unwanted additional spam solicitations. It might result in the sale of your personal computer data to others that might wish to spy on your electronic activities. Any attachment from an unknown source can be opening the door to a dangerous virus. Such malware can certainly do a great deal of harm. For many years such malware has been circulated that attacks all the files on one's computer and utterly destroys them or prevents access to them.

You are in danger unless your computer is protected by a high quality software shield and a firewall designed to protect against viruses. Fortunately there are antivirus software programs out there to help. These include anti-virus software such as those by *McAfee, Symatec's Norton Antivirus,* or other computer security software that we will introduce in greater detail in Chap. 2. But there is now even more scurrilous malware out there.

## CryptoWall 2.0

One of these malware cyber-attack systems is called CryptoWall 2.0, which is a form of software known as a Trojan horse (or simply Trojan). This is one of several new "immuno-resistant viruses" that can invade your computer and seal off all of your files. CryptoWall, CoinVault, and CrytoLocker are file-encrypting ransomware created to blackmail victims into paying to get their files back after a ransomware attack. As the victim you will receive a message that demands payment of perhaps $500 to re-achieve access to the thousands of files and installed programs on your computer and all your applications programs as well. This you are asked to pay in a short period of time—like a day or 2. There may well be a premium of $1000 if you do not pay by the deadline. And then there is the total destruction of all of your files if you opt not to pay at all. If you pay, you may or may not get a security code that can be used to unlock access to your computer files and programs.

The threat posed by CryptoWall 2.0 and other ransomware malware of its ilk is something that creeps out onto "blackhat" websites and starts in one location and then spreads around the world. According to reports by Norton Symantec the use of what is a particularly nasty type of Trojan horse ransomware started in the United States and is spreading to Europe and other parts of the world. Monitoring during 2014 showed this type of growth pattern [8] (Fig. 1.2).

## The Trade in Swipeable Credit Card Information

If you have credit cards with a strip on the back for swiping at the checkout counter you are vulnerable as well. If you live in Europe you are much safer in that the banks have been issuing cards with smart chips that are much

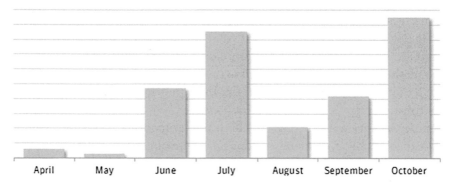

| April | May | June | July | August | September | October |

**Fig. 1.2** Blossoming spread of CryptoWall 2.0 ransomware in the second half of 2014 (Graphic courtesy of Norton-Symantec.)

**Fig. 1.3** A typical looking "smart chip" EMV credit or debit card

more difficult to counterfeit. These chips in your card, sometimes called EMV chips—for Europay, MasterCard, Visa—are much more secure than a swipe card. The EMV name thus comes from the three credit card banking systems that first developed them in Europe and which have been in use in Europe for a decade (Fig. 1.3).

Let's say that you had a credit card with Home Depot, Target, Staples, or other retailer whose credit card data files have been hacked. What is the danger if you are using a simple swipe card? The danger occurs when a large number of credit card information is stolen. (Target files were stolen over a series of months, and an estimated 110 million customer files were obtained by the cyber-crooks.) After this massive theft, this stolen data is actually "sold" on so-called black hat sites. The illicit sale to the next tier down might allow cyber-crooks to get data files for 100 credit cards at a bargain price of just $500. It is then the "retail" cyber-crooks who create bogus cards with the data embedded in the swipe strip of the counterfeit cards. They then hit these accounts fast across many ATMs and retail purchase locations. This might be for ATM withdrawal of say $50–$100 for each card before they are discarded or perhaps resold to less sophisticated cyber-crooks even further down the food chain.

The $500 purchase price and the minor cost of manufacturing bogus credit cards—that is, once you are fully set up for this type cybercrime activity—may allow cyber-crooks a net profit of $5000–$10,000 in just a day or two of criminal effort. This is the return even after deducting all costs and regardless of whether some of the cards have been reported stolen and cannot be used. Even if five of the counterfeit cards get eaten at the ATM machine and some of the cards had very low limits, the return on investment is very good. The real profit, however, comes to those that steal 110 million records and then sell them "retail" at perhaps $5 a pop, and thus realize hundreds of millions of dollars of illicit gain over time.

A Cambridge, Massachusetts, based security firm named BitSight Technologies recently did an assessment of 300 large retail companies to see how well they were protected. The result was not good. BitSight Technology concluded that 58 % of the retailers they assessed had lost ground in terms of being protected from "black hat" hackers. This was not because retailers had lessened their protective shields against hackers. The reduced level of security was because hackers now had access to sophisticated software that was more adept at penetrating firewalls and stealing data [3].

In short the black hats are winning. They now have programs that are faster and more successful in penetrating "corporate firewalls" and gaining access to your credit card information. People think the problem is having your credit card stolen, but the greater danger is having your records stolen, along with millions of other people.

## *The Cyber Criminal Bazaar*

The well-known technical consulting firm The Rand Corporation released a detailed report on an alarming trend in cyber-criminal activity—the creation of a black market network of "black hat" hackers that cooperate on what is now an international scale. These cyber-criminal marketplaces perform a multitude of functions. The illicit online networks provide tools known as exploit kits for illegal monitoring and hacking into user Internet connections. They sell credit card information and virtually offer seminars in how to commit cyber-crime.

This report suggests that these trends will continue with more activities being conducted within darknets, the rise and expanded use of crypto-currencies, the development of greater anonymity capabilities in malware, and more attention to encrypting and protecting communications and transactions by an international network of cyber criminals.

Based on current trends, this report predicts that there will be: (1) a wider range of opportunities for black markets; (2) more hacking for hire; (3) more diverse types of hacker support service offerings; and (4) brokers for stolen data and information that range from financial records and credit card infor-mation to various types of intellectual property and even espionage intelli-gence. These organized cyber-criminal networks will be equipped to carry out all forms of cyber-attacks that include persistent assaults, targeted attacks; opportunistic cyber thefts, mass "smash-and-grab" attacks and more [9].

The following 11 key findings from the Rand Report include:

1. The hacking community and cyber black markets are growing and maturing.
2. The cyber black market has evolved from a varied landscape of discrete, ad hoc individuals into a network of highly organized groups, often connected with traditional crime groups (e.g., drug cartels, mafias, ter-rorist cells) and nation-states.
3. The cyber black market does not differ much from a traditional market or other typical criminal enterprises; participants communicate through various channels, place their orders, and get products.
4. The evolution of the black market mirrors the normal evolution of markets with both innovation and growth.
5. The cyber black market can be more profitable than the illegal drug trade.
6. Cyber black markets respond to outside forces.
7. Because of an increase in recent takedowns, more transactions have been moving to darknets; stronger vetting is now taking place; and greater encryption, obfuscation, and anonymization techniques are being employed, restricting access to the most sophisticated parts of the black market.
8. The proliferation of "as-a-service" and "point-and-click" interfaces low-ers the cost to enter the cyber black market.
9. Law enforcement efforts are improving as more individuals are techno-logically savvy; suspects are going after bigger targets and thus are attracting more attention; and more crimes involve a digital compo-nent, giving law enforcement more opportunities to catch crime in cyberspace.

10. Still, the cyber black market remains resilient and is growing at an accelerated pace, continually getting more creative and innovative as defenses get stronger, law enforcement gets more sophisticated, and new exploitable technologies and connections appear in the world.
11. Products can be highly customized, and players tend to be extremely specialized [10].

These developments can be thought of as both good news and bad news. The good news is that law enforcement has now recognized that cyber criminals are often linked to a well-organized, frequently international, and networked group of individuals hiding behind encrypted networks. As can be seen in the official cyber security programs being undertaken in the United States, Europe, Japan, and OECD countries, as contained in Appendices B, C and D, these countries have stepped up their game. Such programs involve more international cooperation, specialized training, and more sophisticated tracking and diagnostic tools to bring organized cyber-criminal and even techno-terrorists to justice.

The bad news is that what were once one-off actors acting alone have the support of an international fraternity connected by darknets that represents a sophisticated and specialized network providing a wide range of services. The most worrisome thing of all is the presumed ability of techno-terrorists seeking to inflict massive Level Three attacks against nations linking up with these internationally linked cyber-criminal bazaars (Fig. 1.4).

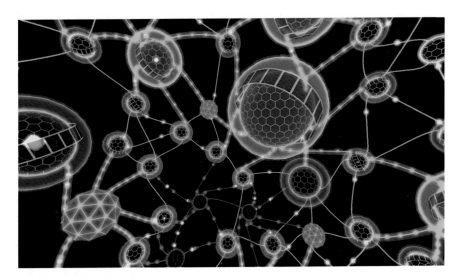

**Fig. 1.4** Darknets that are heavily encrypted and containing hidden connections are increasingly how cyber criminals communicate

## Cyber Thieves and Medical Records

In February 2015 Anthem Blue Cross, the second largest provider of medical insurance in the United States, revealed that black hat hackers had obtained records of 80 million customers. Suspicions were subsequently thrown on Chinese black hat hackers. Anthem reported that while financial data might have been revealed, the medical records had not been compromised. This is significant in that while stolen credit cards sell for perhaps $5 each, stolen medical records sell for more—as much as $25 to even $100 each, depending on how many details are revealed. Cyber-crooks can seek bogus payments for major medical bills and operations and purchase expensive medical equipment such as electrical scooters or oxygen tank systems that can be sold for large amounts, even if highly discounted. In short, if cybercriminals get access to detailed medical records, they can make thousands of dollars in false claims and pocket the money before it has even been discovered that the records were stolen.

Efforts to block credit card fraud by the introduction of chips in the latest rendition of smart cards will rather logically shift cybercriminals' attention to other prey. The most logical alternative will undoubtedly be cyber-attacks on medical records and efforts to divert financial funds by capturing key information that allows false claims to be made [11].

## Nigerian Prince Come-Ons and the Latest in Targeted Online Scams

There are also the bogus solicitations that come to you in e-mail. The clumsy requests from "Nigerian princes" for you to send them a small amount of money so you can collect millions are now well known. Today's solicitations, however, are much more sophisticated. Hackers can go to Facebook or other social media pages and collect a very large amount of information about people. People become vulnerable when cyber-crooks go online and find out who your friends are, where you live, the names of children and their ages, and where you golf or play tennis or swim, and even where your children go to school or their friends' names. This allows them to send a message to you seeking emergency help that seems to be coming from a close friend or relative that is stranded on a trip abroad. It could seem to be coming from your bank and broker seeking a change to your account and its password or something else entirely. The purpose, however, will be to seek a money transfer, the revealing of one or more of your key passwords, or some other way to relieve you of your money.

## Protecting Against Personal Cyber Attacks

The above examples are only a few types of cyber-attacks that can be made against you and your family. In some ways these are the most benign. If you are someone of means, then a review of your Facebook account or other social media might be used to plan a successful kidnapping of you or a loved one. Or perhaps it is someone that is planning a robbery of your house that has hacked your e-mails to find out when you are away and the house is unattended. Cyber criminals might even be able to learn the code to your home security system, if you were ever to e-mail it to someone tending your home while you were going away.

There are ways to protect against such cyber-attacks, and we will address them in detail in Chap. 2. The main rule of thumb is to exercise good judgment. Back up your data on an external memory source such as an external hard drive that is not connected to your computer. Don't post anything on your website or social media that could be sensitive or used by a cyber-crook. Protect your passwords, Social Security numbers, and other personal information. In a number of cases there are emerging better protective strategies that just a password. For instance, any smart phones today offer facial recognition or drawing of designs or various forms of biometrics that are much more secure than passwords [12]. There are even low cost services that can be obtained to assist with password protection as well as to help with data backup in addition to protecting against malware such as viruses and worms.

## Phishing and Pharming

Phishing is one of the most common categories of online scams. This is where a criminal, typically by means of high volume spam e-mails and/or the establishment of fake websites set up to appear to be legitimate, convinces victims to provide personal information. The sought-after information might be such data as private account details, credit card numbers, and/or Social Security numbers. There are more specialized forms of phishing that are targeted to particular individuals. These are variously called "spear phishing" (i.e., a cyber-attack on selected individuals), "whale phishing" (i.e., going after a key executive or person of great influence), and "clone phishing" (i.e., getting information from someone's associate and then attacking a specific individual using the associate's or friend's faked identity). These and other terms are described in greater detail in the glossary of terms.

Pharming is a similar type of cyber-attack to phishing, where the attacker tricks their victims into providing personal information to a malicious website. The mechanics of how this works is explained in more detail in the glossary. The main thing to know is that Internet domain names can be tampered with and so even if a message or web site looks like it is coming from a legitimate bank, institution or even the U.S. Government this may not be the case.

# Level Two Cyber-Attacks

You should also be alert to attacks against banks, retailers with credit cards, or even businesses where you have an account. The biggest threat you have to worry about is that the attack could involve your personal accounts.

## *Trojans on the Attack Against Banks*

There are various types of attacks known as Trojan horses, data bombs, and other malware that could do your bank, brokerage house or other type of credit card issuers a good deal of harm. An attack on an individual via a Trojan horse-type malware is one thing, and a $500 loss is certainly uncomfortable. A full-scale Trojan horse attack on a major bank is a horse of quite a different color. This type of attack is much more sophisticated, and would seek to create an undetected virtual digital pathway into the bank accounts that record all banking or financial operations, perhaps for many weeks.

When the Trojan horse is withdrawn by the hacking cyber-crooks the banking records for a several week period could suddenly disappear. In this type of Trojan horse attack, the magnitude might involve perhaps billions of banking records. The key question then becomes: What is the intent of the attack? The most typical case would be very large-scale ransoming whereby the bank or stock brokerage quietly pays handsomely to get their records back, and then no one is the wiser. The ransom is written off as a business loss. Also the person in charge of IT security probably gets fired, and new encryption and telecom systems are installed as well. It could be a cyber-terrorist attack, and the bank could be laid to ruins, unless there is a backup system for the data every day, and only 1 day's transaction data are lost in the attack.

Regardless of the intent, you could become a collaterally damaged account holder in the bank. Keeping a current paper copy of your records is always a good idea. Diversification is also a good idea. Don't keep all of your bread (the green kind) in a single breadbasket (i.e., bank or brokerage house). A Trojan horse could, perhaps even by accident, gobble up your records. In short, a number of things could go wrong, and your assets along with thousands of others, could indeed be at risk. Unless you have identity theft insurance, you might be in for a substantial loss. Fortunately a number of protective cyber services offer up to $1 million in identity theft protection [13].

## Insurance Collapse Due to Massive Cyber Attacks

Another industrial sector that is at increasing risk to cyber-attack is the global insurance industry. Large multinational insurance companies insure trillions of dollars of assets around the world. The occurrence of natural disasters such as hurricanes or typhoons can create major claims mounting into the billions of dollars. Fortunately insurance companies typically have enough assets to pay off claims and still remain solvent. A cyber-attack of sufficient magnitude that it takes out a nuclear power plant, or poisons an entire water supply, or destroys an entire city, could bankrupt an entire insurance industry. Thus cyber-attacks on banks can cripple your accounts. A devastating cyber-attack that is sufficient to destroy cities can also bankrupt your insurance carriers. The truth is that while we now have stress tests for banks and the quality of their assets we do not have any way of assessing the survivability of the insurance industry in the advent of a cosmic disaster (i.e., a giant asteroid strike or solar coronal mass ejection) or a monumental cyber-attack that creates a nuclear meltdown or other horrendous event involving millions of people.

A number of space-faring nations, such as the United States and France, have passed legislation to provide a cap on the liability associated with damages that might occur in the course of a launch such as a rocket exploding in Miami, Florida. It might be appropriate to consider putting a cap on insurance claims in the case of a catastrophic natural disaster or a cyber-attack that creates massive damages to property and human life [14].

### The "Internet of Things": A World in Which Cars, Appliances, and Almost Everything Is Connected to the Net

The Internet of Things (IoT) involves all digital devices, including household appliances and even electronic components connected to the Internet. A recent study by Cisco concluded that there might be as many as 50 billion electronic devices connected to the Internet by 2020. A Research and Markets study estimated that over the next 5 years, there will be tremendous growth for machine-to-machine (M2M) communications fueled by the connectivity demands of the Internet of Things. This implies a projected global growth of annual M2M IT services from $260 billion to $1 trillion dollars. The problem is that smart appliances such as washing machines or refrigerators can now be commandeered and hacked. In the "IoT" world, microprocessors can be reprogrammed for a variety of purposes, ranging from sending out thousands of spam messages over the Internet to initiating untraceable messages that lead to serious cybercriminal or even techno-terrorist attacks (Fig. 1.5).

**Fig. 1.5** The "smart" world of the Internet of Things, where even vacuum cleaners and lights are connected to the Net

# Level Three Attacks

The loss of your bank account or some other form of financial loss is definitely a serious threat. The attack on countries and vital national infrastructure is even more threatening. Today, the world around you is vulnerable to attack through a morass of information networks. The water you drink, the food you eat, the subway or train or plane you ride in, the power systems you depend on, and indeed every vital infrastructure in your life is now dependent on intelligent information and telecommunications systems. As we make the transition to a world controlled by information stored in the Cloud and the so-called "internet of things" our potential vulnerability will unfortunately increase. A few examples will illustrate the point.

## *GPS Resilience and Vulnerability*

A recent report by the U. K. Royal Academy of Technology on GPS reliability and vulnerability concluded that 8 % of the economy of Europe is dependent on Positioning Navigation and Timing (PNT) satellite services. The likelihood is that the dependence of the U. S. economy is even higher. It turns out that military planes, UAV drones, Internet synchronization, and a host of military systems are also dependent on the GPS satellite network. GPS jamming, a severe solar event, an EMP explosion, or another type of accident or terrorist attack might bring down the GPS system. This would have a negative impact on your life in more ways than disabling the navigation system in your car. The loss of synchronization of the Internet alone would create major economic repercussions. A number of people today are almost totally dependent on their smart phones for communications, contact information, navigation, appointments, and more. The complete dependence on IT systems might make you more vulnerable than you know. It is prudent to store in a safe or safety deposit box originals of birth certificates, marriage licenses, printouts of bank records and brokerage accounts, Xeroxes of your passport, a printout of personal contacts (address, telephone number, e-mail) and stock certificates. You need to think about ways that you could live your life if you needed to live it without a computer, a smart phone and GPS signals. Too often that which seems impossible to happen in life can actually come true.

We have become incredibly dependent on what might be called our intellectual prosthetics. Going cold turkey on our electronic servants is increasingly difficult to cope with. Such overdependence makes us vulnerable in many ways that we will explore in coming chapters. Unless you have a generator that runs off of a bicycle pedaling device or a large battery complex rechargeable by the Sun, your various communications and cyber devices would become useless if you and your community were to lose electrical power for a sustained period of time.

## Attack on Water and Sewage Systems via SCADA Systems

Have you ever wondered how our traffic signals inside our communities change their timing to adjust to rush hour? Or how the flow of oil through national pipelines or water or sewage is routed to proper destinations at the right speeds and pressure levels? An incredible network of control systems is at work across the United States and around the world. These computer-driven SCADA (Supervisory Control and Data Acquisition) networks replace millions of workers. These "smart systems" work 24/7 to route and control electric power flows, distribution of oil, natural gas, water, sewage, control of traffic lights, operation of building controls, elevators and escalators, and more.

Such digital remote control computer systems indeed supervise the opening and closing of valves, the routing of vital resources, and the provision of billions of instructions, and collection of even more data to monitor the critical infrastructure. Unfortunately the security on these SCADA networks can be inadequate. Water and sewage systems sometimes retain the same security control codes that came from the manufacturer without change.

The pervasiveness of SCADA networks in modern society and their potential for being hacked via criminal and terrorist attack is of truly major concern. One of the many dangers that might possibly occur with devastating effects on public safety would be where a terrorist diverts sewage flows into municipal water supplies. This might be one of the easiest types of SCADA-type attacks to mount, but controls on traffic signals, elevators, electrical power transformers, or oil pipelines might allow a techno-terrorist to bring destruction to our country and perhaps your community at an even larger scale [15].

## *Transit Systems and Aircraft Safety at Risk*

Even more ominous would be major cyber-terrorist attacks on civil or military aircraft, or a national train system. These facilities are run by human traffic controllers. Nevertheless, the extreme complexity of these systems still require computer software and networking in order to operate. Spurious software could sabotage the U. S. civil air traffic control system, which coordinates up to 20,000–25,000 planes in the air at any particular time. The train scheduling system is likewise complex and driven by computer-based algorithms. A country that wishes to attack another has a wide variety of options open to them, but as shown by the example of the attack on Sony, such a cyber-attack is difficult to hide. The key thing to grasp is that a cyber-attack could escalate to a level that equates to or even exceed a nuclear missile attack. And what makes the threat so very dangerous is that the general public rejects the idea that this could be true.

# A Recap of Why You Should Care

The reason why you should care about cybersecurity is that you are increasingly vulnerable to a wide range of cyber-attacks. The five main risk elements are the expanded availability of Wi-Fi systems; automatically accessible features (i.e., vulnerable apps) on smart phones; "the Cloud"; expanded forms of malware; and now the Internet of Things. These cyber vulnerabilities are exponentially increasing one's personal risk of cybercriminal assaults. There are defensive strategies, but it is difficult to stay ahead of the curve, and the Internet of Things (IoT) may prove to be the most difficult of the cyber-challenges in the decade ahead. It is clear what the advantages are to vendors when it comes to the Internet of Things, but it is really not as clear what the advantages are to consumers. In many ways the Internet of Things is like inviting a spy into your home or office.

These various cyber vulnerabilities can threaten your bank or brokerage account, your credit rating, expose you to loans you did not make, and a host of other financial losses. It can expose you to identity theft. Identity theft may involve more than a loss of money. You could actually end up with you being charged with a misdemeanor or even a felony such as grand theft or embezzlement. These cyber vulnerabilities are all in the context of so-called Level One attacks.

Then there are also Level Two attacks (i.e., assaults against banks, brokerages, insurance companies, and even movie studios) that could also bring you grief as collateral damage. These types of attacks could range from minor inconveniences up to finding that your insurance carrier or brokerage firm is bankrupt, the failure of a business on which you depend for employment, a breakdown in transportation, or the unavailability of critical medicines.

Finally Level Three cyber-attacks are at some levels the most to be feared. There is a much greater danger of a national or organized terrorist attack using cyber malware than there is of a nuclear missile strike or the explosion of a "dirty bomb." This is why the U. S. Cyber Command and similar efforts to create cyber defense systems that are occurring in 60 countries around the world have now become critical. A Level Three cyber-attack might be undertaken by one country versus another, but it is far more likely that it would be a terrorist entity such as ISIS (or ISIL). It is far easier for an ISIS or an Al Qaeda cell to mount a techno-terrorist attack against the United States or another country with a heavy reliance on computer networks than an armed attack. Such a large-scale attack might conceivably crash planes in the skies, derail trains or blow up a nuclear power plant and do so from a remote location an ocean away. Even computer experts may dispute that such things are possible. Yet there is strong evidence that horrendous computer attacks are possible in the United States and truly in most countries in the world.

We hope you take these cyber threats seriously. Cybercriminals, technoterrorists and cyberbullies are unfortunately poised to attack you and your loved ones on a relentless 24/7 basis. Read on to learn how you might defend yourself with the tools and techniques that are available. There are a growing number of these and software packages that allow you to defend against cyber-attacks, but unfortunately there are also a growing number of tools available to cyber-crooks, who can use them to penetrate your cyber-defenses. With the existence of the "darknet," the ability to spread these cyber-attack tools faster and farther is growing. We also recommend that you support national legislative reforms that assist in protecting institutions and nations against vicious and potentially disastrous financial cyber-attacks on your computer and smart phone systems. Today such cyber-protections are indeed part of the war on terrorism as well as the war on crime. In July 2015 the U.S. Government working with 19 other government shut down "Darkode" one of the largest and most comprehensive darknets. That was the good news. The fact that there are an estimated 800 such sites out there is what gives great pause to anyone who is concerned about digital attacks on their privacy and online assets [16].

# References

1. David Raven, "Nanny terrified after man hacks baby monitor and calls little girl "cute" over speaker." The Daily Mirror, January 28, 2015. http://www.mirror.co.uk/news/world-news/nanny-terrified-after-man-hacks-5056098
2. Southsea Man Helps Steal £22,000 from Woman's Bank Account After her E-mail is Hacked. http://www.southnews.org.uk/news/southsea-man-helps-steal-22000-from-womans-bank-account-after-her-e-mail-is-hacked.
3. Top Ten Identity Thefts, U.S. Internal Revenue Service. http://www.irs.gov/uac/Newsroom/IRS%E2%80%99s-Top-Ten-Identity-Theft-Prosecutions.
4. A Parents' Guide to Cybersecurity, www.ConnectSafely.org/security.
5. "Mum turns troll-hunter and forces 500 cyberbullies off the internet" The Mirror, August 18, 2013. http://www.mirror.co.uk/news/uk-news/mum-turns-troll-hunter-forces-500-2180859
6. "Andrea Peterson, "How easy is it to hack an airplane?" Washington Post, April 21, 2015, p. A-14.
7. "Techopedia explains cyberattack." http://www.techopedia.com/definition/24748/cyberattack
8. Susan Gallagher, "Inside CryptoWall 2.0: Ransomware, professional edition" ArsTechnica, Jan. 9, 2015., http://arstechnica.com/information-technology/2015/01/inside-cryptowall-2-0-ransomware-professional-edition.
9. Lillian Ablon, Martin C. Libicki, Andrea A. Golay, "Markets for Cybercrime Tools and Stolen Data: The Hackers' Bazaar" Rand Corporation Report, http://www.rand.org/pubs/research_reports/RR610.html (2014).
10. Ibid.
11. Somerville, Heather, "Retailers' data breaches could get 'ugly' during holiday season" Mercury News, Nov. 27, 2014. http://www.mercurynews.com/business/ci_27025645/retailers-data-breaches-could-get-ugly-during-holiday.
12. "How To Make Passwords Obsolete" Information Week, April 7, 2015. http://www.informationweek.com/cloud/how-to-make-passwords-obsolete/d/d-id/1319775?mc=NL_IWK_EDT_IWK_daily_20150408&cid=NL_IWK_EDT_IWK_daily_20150408.
13. "Pharming and Phishing Definitions." www.techterms.com/definition/pharming.
14. Space Launch Liability Indemnification Extension Act. U. S law that assumes liability from $500 million to $2.7 billion in the case of a launch disaster. Private insurance can be obtained for launch insurance up to $500 million.
15. Joseph N. Pelton, Indu Singh, ElenaSitnikov, "Cyber Threats, Extreme Solar Events and EMPs." Inside Homeland Security, March 2015.
16. Ellen Nakashima, "U.S., 19 other nations crush criminal hacking ring" Washington Post, July 16, 2016, p. A4.

# 2

# Where to Go for Assistance

## Help Available

Being secure in your use of the Internet in today's world is difficult, but there are a number of services out there that can help. In the last few years the range of electronic services available to consumers has mushroomed, and this is a great boon in many ways. Many goods and services are available at lower costs, and the range of options has increased. Unfortunately, expanded opportunities available via the Net come at a hidden cost. That cost is exposure to cybercrime and unwanted spam messages. If you have children there is also the risk of them being exposed to pornography or inappropriate information.

In the world of the Middle Ages, information was a scarcity. There were very few books, and they were all handcrafted. The cost of a single book was huge, and only the elite could access most of the world's knowledge. Over the centuries information has become more and more accessible, and the cost of obtaining information has plummeted. It was said in the 1980s by Richard Solomon of the MIT Media Labs: "Someday in the not too distant future we may be paying more to screen out unwanted information than we pay to get the information we want."

In a world of increasing information overload Richard Solomon's prediction seems to have been incredibly insightful. Indeed we have increasingly been forced to purchase anti-virus filters and firewalls to screen out malware and install spam filters to free ourselves of unwanted solicitations. If one looks at today's world of information retrieval in this context, then the equation becomes clear. We are paying less for most of the information that we receive, and the corollary is that we are now paying for what can be generally characterized as cybersecurity. Thus the still-modest costs of filters

© Springer International Publishing Switzerland 2015
J.N. Pelton, I.B. Singh, *Digital Defense*, DOI 10.1007/978-3-319-19953-5_2

and protective services to keep out unwanted information is just the cost of getting information in today's world.

Today's millennials—let's say people under 30—in general don't subscribe to newspapers or listen to the 7 or 11 o'clock news but simply get their news online. They can purchase much of what they want or need online. The cost of electronic filters to protect against cybercrime and identity theft is on balance less than a newspaper subscription. The devil's bargain is that if you want cheap information off the Internet you have to pay for it by buying some form of cybersecurity and getting professional assistance to help protect against identity theft, ransomware extortion, or other forms of electronic intrusions into your life.

Oddly enough, the easier job that you might undertake to protect you and your family might be in buying and installing various forms of cyber protection that you can buy online, such as we will describe to you in the following pages. For an amount of about $60–$75 a year, if you invest wisely, you can buy cyber protection that includes antivirus, a good firewall, identity theft protection and insurance. You can also learn a number of things to look out for that might lure you into a cyber-correspondence that might lead to theft of your assets or problems with your computer. The number of phishing and pharming scams out there is only going up despite more cybersecurity enforcement. This is in large part due to the fact that black hat hackers come from many countries where enforcement policies are lax.

The bigger challenge may come in trying to protect you and your family against cyberbullying by so-called cyber "trolls" that stir up mischief or worse. Along with these cyber-attacks against your family, and particularly teens and tweens, there are problems associated with pornography on the Net and the temptation of so-called "sexting" (i.e., sending nude or compromising pictures of oneself or others over the Internet). Further there are a lot of hate e-mails out there that come in the form of racial, religious or sexual orientation bigotry or terrorist-generated messages that can be hateful to members of your family or could provide the wrong type of influence on your family and loved one.

The first part of Chap. 2 outlines various services that are available to you to protect you against cyber-criminal attacks that could disable your computer, lead to substantial financial losses, or other nefarious results. These are the most straightforward safeguards we advise you to use. These are forms of protection that can help you defend your computer, cell phone, and other electronic devices against cyber-attacks. Later in Chap. 5 we return to address security for cell phone and instant payment programs such as "Apple Pay." Also in Chap. 2 we discuss some of the services that are available to address and assist with the problem of cyberbullying and cyberbullying alert services.

The last half of Chap. 2 shifts from a discussion of security measures such as anti-viruses and firewalls, to the problem of unwanted content and messages that can make the lives of you and your loved ones miserable. Here, as we noted in earlier, comes the hard part. This is because we are trying to block, mitigate or stop unwanted and unkind content from flowing to you and your family and stopping trolls from polluting the Internet with untrue and hurtful messages about your loved ones. This is difficult and challenging to deal with but also an area where protective actions are possible and the most vicious and harmful "trolls" can be shut down.

First let's protect your computer against cyber-attacks. We have already outlined some of the common sense steps you should take to protect yourself and your records. Now let's examine the type of assistance you might wish to obtain.

## Electronic Filters That Protect Against Malware Intrusions

The first thing you should do is have a good Internet antivirus software package. Here, for instance, is what Norton Management, among others, claims to provide. We provide this list because it is a comprehensive statement and is typical of what a antivirus service should provide. We would like to emphasize that in this book that seeks to provide objective advice, the specific examples provided in this chapter or elsewhere about service provider offerings is not an endorsement of these products. We suggest you go to the consumer rating sites to make your individual choices [1] (Table 2.1).

For about $60–$75 per year or less one can purchase and install one of the top cybersecurity software packages offered by McAfee (now owned by Intel Security), Kaspersky, Bull Guard, ESET, Norton (owned by Sematec), or any of the other options listed on below. The higher prices can include protection for multiple devices (Norton covers up to five or even ten devices for instance at an additional charge) as well as security backup protection. This backup protection is actually quite important and useful. If you have a number of important documents and photographs that you consider vital, it is important to have automatic backup of this information against a cyber-attack. Norton service claims in Table 2.1 are used only as an example and not as a product endorsement.

Table 2.1  What Norton claims its software products can do

| Key features |
| --- |
| *Actively Protects Against Viruses, Spam, Identity Theft, and Social Media Dangers* |
| • Insight identifies which files and applications are safe and which are dangerous, using the combined feedback of more than 175 million Norton users. |
| • Norton Community Watch tracks virtually every file on the Internet for comprehensive global threat monitoring. |
| • SONAR Behavioral Protection detects signs that a file is dangerous to proactively protect you from never-before-seen threats. |
| • Spam blocking keeps your mailbox free of unwanted, dangerous, and fraudulent emails. |
| • Internet Protection System scours websites and social networking sites for suspicious links and content to identify the latest social networking scams. |
| • Download Insight and IP Address Insight to prevent you from downloading files from websites that have a low reputation score within the Norton user community. |
| • Live 24 x 7 Threat Monitoring is backed by a network of Norton users who serve as your own personal Neighborhood Watch group. |
| • Scam Insight reviews a website's reputation and lets you know if it's safe to enter your personal information. |
| • Anti-phishing technology blocks fraudulent "phishing" sites set up to steal your personal information. |
| • Identity Safe remembers, secures, and automatically enters your usernames and passwords for you, so they can't be lost or stolen. |
| • Parental control feature helps you protect your children from online dangers by giving you direct access to Norton Online Family. |
| • Safe Web tells you if a website is unsafe before you visit it, and it's too late. |
| • Safe Web for Facebook scans your Facebook wall and news feed for URLs containing security threats such as phishing sites, malicious downloads, and links to unsafe external sites. |
| • Intelligent 2-way Firewall prevents strangers from accessing your home network by blocking incoming traffic determined to be unsafe. |
| • Network mapping and monitoring shows all the devices connected to your home network, so you can spot uninvited guests using your wireless connection and/or eavesdropping on you. |
| • Automatic silent updates keep you one step ahead. |
| • Automatically downloading of updated products and installing important product and feature updates when you're not using your computer. |
| • Norton Pulse updates virus definitions every 5–15 min without disrupting work or play. |
| • Insight + Optimized File Copy identifies safe files and only scans unknown files. |
| • Built-in Intelligence maximizes battery life by putting off non-critical activities until you are plugged in and out of full-screen mode. (Table provided by Norton -by Sematec) |
| • Norton Management enables easy single-password access to all Norton Cloud-based applications and web properties |

Matchtop's Top Ten Antivirus Sites evaluation (see below) of what it considers the best programs out there, based on levels of protection and cost, uses a 5.0 scale, with 5 being the highest rating [2] (Table 2.2).

Table 2.2  One assessment of various antivirus software commercially available

| Rating the top ten antivirus sites for performance and cost[a] | | |
| --- | --- | --- |
| Company | Rating[a] | Cost/year |
| McAfee by Intel Security | 4.9 | $24.95 |
| KasperSky | 4.5 | $29.95 |
| Bull Guard | 4.5 | $23.95 |
| ESET | 4.5 | $20.15 |
| Norton by Sematech | 4.4 | $35.99 (for up to five devices) |
| Trend | 4.4 | $29.95 |
| AVG by Microsoft | 4.4 | $31.99 |
| Zone Alarm | 4.3 | $44.95 |
| VIPRE | 4.2 | $39.95 |
| PANDA | 4.1 | $39.95 |

[a]See the website address given below for details about the ratings

It is recommended that you explore all of the three sites indicated below. These websites provide a very useful assessment of various antivirus materials that are currently on offer. Some are better at providing instant updates. Others offer better protection against phishing or certain types of malware. The main thing in making your decision is that there are number of websites that can provide useful advice. In addition to the three websites provided below, Consumer Reports and the AARP websites are also quite helpful. One of the major things that should be considered in making your choice of a cybersecurity package is how many devices do you wish to protect? If you protect your home computer but not your smart phone or laptop or netbook or your IPad, Kindle, Nook or tablet, you could be in for big trouble. The issue of mobile services and cybersecurity we address in Chap. 5.

## *Recommended Websites*

http://antivirus.thetop10sites.com/
http://www.top10antivirussoftware.com/
http://www.pcmag.com/article2/0,2817,2388652,00.asp

It should be recognized that the above list of ten is still not comprehensive, and that there are other products out there such as "Avast", "Aviro", "BitDefender" and "Pareto" plus others that should be seriously considered. This is particularly the case with Avast and Avira because their basic level of protection is for free. If you are going to focus on just free services, we suggest that you go to the *PC Magazine* site (http://www.pcmag.com/article2/0,2817,2388652,00.asp) that does an independent evaluation.

Indeed there are several cybersecurity products out there that allow "free" antivirus downloads. These include offerings from organizations such as Avast.com or Avira.com. These organizations provide free downloadable antiviruses, at least for their lowest level of protection—and sometimes the antivirus is only free for just a year. Of course, higher levels of protection are available for a fee. These groups' marketing strategy is to offer free service as an entry point to sell their additional software. Also, if you purchase Zone Alarm or Comodo Firewall protection, the company will provide free antivirus with its product.

Thus, for $35–$60 a year you can obtain a reasonable level of protection against viruses, worms, Trojan horses, spyware, ransomware, and keylogging. These terms are all defined in the glossary, but the main thing to know is that these are all bad things that black hat hackers, or "crackers," can do to steal information from you while you are logged on to the Internet or using your smart phone.

The basic question that you must decide is which service to use. You might consult such trusted websites as Consumer Reports or AARP, but there is no specific "right" choice. The definitely wrong answer is to say I will take my chances and not get at least basic antivirus protection. If your funds are very limited at least opt to sign on with AVAST or Aviro.

## Identity Theft Protection

The biggest nightmare that a person might encounter in today's cyber-world is identity theft. This means that someone not only takes over your identity but somehow manages to wipe out your bank account, switch your social security payments, or take out a large loan in your name for which you could be legally responsible unless you can prove the identity theft in court. Such identity theft can, and often does, occur online, but it can also happen by someone that targets your regular mail, your discarded but not shredded financial records, or by other means of eavesdropping. The most common means of identity threat, however, occurs online by intercepting passwords and accessing financial records via a keylogging program.

One might think that these are just isolated incidents that happen to only a few people, but a study conducted by Javelin Strategy and Research concluded that 11.6 million Americans had some sort of identity theft issue during the course of a year. Other studies have projected different numbers of incidents, but clearly this is a widespread problem in the United States and around the world [3].

Fortunately protection is available against identity threat—and at reasonably low rates. These services typically cost between $125 and $330 a year and usually provide a $1 million warranty against financial loss due to identity theft. Below is an analysis of ten sites that provide identity theft protection and $1 million in insurance coverage. For more details, go to www. top10identitytheftprotection.com [4] (Table 2.3).

If you do sign up for identity theft protection we believe that the anti-keylogging/antivirus service is important to include, unless you already have this capability installed in your computer devices. The AARP service via Equifax is perhaps the lowest cost offering available, especially to seniors who are AARP members, but it is recommended that if one takes this service they have an antivirus service that protects against Internet hackers.

The types of services that can come with identity theft protection services varies a great deal. Lifelock, which actively markets its services via extensive television ads, offers services that range from $9.95 a month up to $29.95 a month. The range of services that are available is shown in the table below. Note, before you sign up, you should check carefully the range of services listed above, since these are given by the providers. Some of the providers have been sued for overstating their levels of protection. The chart below indicates a range of protective services offered by LifeLock that range from "Standard" to "Ultimate." Other providers offer alternative levels of protection. It is the opinion of the authors that most individual consumers that don't own businesses do not need to purchase higher levels of protection, but feel the $1 million service guarantee is key. Again, we are not endorsing LifeLock but merely citing its tiered level of protection as not untypical of identity theft protection services [5] (Table 2.4).

In making your actual decision about which service to select we suggest that you look at objective third parties in making your choice. Remember that those that advertise the most are not necessarily the best or cost effective. Three sites that you might consult in making your decision include:

http://www.consumeraffairs.com/privacy
http://www.thegeekprofessor.com/
http://krebsonsecurity.com/2014/03/are-credit-monitoring-services-worth-it/

Table 2.3 Identity theft services[a]

| Service name/rating | Price | Fraud monitoring | ID theft insurance | Reports delivered | Computer security | Bottom line |
|---|---|---|---|---|---|---|
| Identity Guard | Free 30-day trial; $14.99/month (after our 25 % discount) | Monitors 3-bureau credit report, credit cards, public records, social security number, applications, Internet security | $1 million | 3-bureau credit scores and a public record report each quarter | ZoneAlarm Internet security suite; anti-key-logging software | Most complete identity theft protection service we reviewed; 3-bureau credit report monitoring; credit report/score updates every quarter; 25 % discount and free 30-day trial |
| Trusted ID (AARP partnership) | Free 14-day trial and 10 % discount; $9.38/month (paid annually) | Monitors 3-bureau credit report, credit cards, public records, social security number, bank accounts, medical records | $1 million warranty | Equifax credit reports and scores monthly; TransUnion, Experian credit reports and scores annually | None | Best value, especially for families; full credit report monitoring; monthly Equifax credit reports and scores; 10 % discount and free 14-day trial |
| AARP | Free 14-day trial and special AARP price; $9.17/month (paid annually) | Monitors 3-bureau credit reports, bank accounts, credit cards, social security number, public records | $1 million service warranty | Equifax credit reports and scores monthly; Transunion, Experian credit reports and scores annually | None | Comprehensive identity theft protection and credit report monitoring for AARP members and family; monthly Equifax credit reports and scores; special AARP price and free 14-day trial |
| Lifelock Ultimate | Free[a] 30-day trial; $24.75/month (w/annual prepay and 10 % discount) | Monitors 3-bureau credit reports, applications, credit cards, social security number, driver's license, address change, credit card and bank account activity | $1 million guarantee | TransUnion credit scores monthly; 3-bureau credit reports and scores annually | None | Thorough identity theft protection and 3-bureau credit report monitoring; annual 3-bureau credit reports and scores; monthly TransUnion credit scores; somewhat costly even with 10 % discount; free[a] 30-day trial |

| | Price / trial | Monitors | Insurance / guarantee | Credit reports | Norton Internet Security Online | Notes |
|---|---|---|---|---|---|---|
| Privacy Guard | 30-day trial for $1; $14.99/month | Monitors 3-bureau credit report, Internet security | $1,000,000 insurance | All 3 bureau reports and scores monthly, social security report, medical info bureau report | Norton Internet Security Online | Solid credit protection with monthly credit report/score updates; includes our top-rated Internet security software; 30-day trial for $1 |
| ID Freeze | Free 14-day trial and 15 % discount; $7.01/ month (paid annually) | Monitors credit cards, public records, social security number, bank accounts, medical records | $1 million warranty | All 3-bureau credit reports each year | None | Reasonably priced identity theft protection for individuals and families; doesn't provide credit report monitoring; 10 % discount and free 14-day trial |
| Lifelock | Free[a] 30-day trial; $8.25/month[a] (w/ annual prepay and our 10 % discount) | Monitors credit cards, social security number, driver's license on Internet black market and address change verification | $1 million guarantee | None, unless plan is upgraded | None | Valuable identity theft protection and customer support for an affordable price, yet lacks in terms of credit report monitoring; 10 % discount and free[a] 30-day trial |
| Legal Shield | $24.95/month and one-time $10 membership fee | Monitors 3-bureau credit report, credit cards, emails, phone numbers, Social Security number, bank accounts, medical records | None | All 3 bureau credit reports each year | None | Somewhat pricey when compared to other services; complete restoration assistance; no insurance/guarantee or security software |
| Protect my ID | 7-day trial for $1 with enrollment in ProtectMyID; $19.95/mo | Monitors only Experian credit report, credit cards, new financial accounts or applications, address changes, public records | $1 million | One Experian credit report | None | An expensive option for ID theft protection and lacks in protection; only includes Experian credit report monitoring; 7-day trial for $1 with enrollment in ProtectMyID |

[a]Table prepared by top10identitytheftprotection.com

Table 2.4 Example of typical service provider offering identity theft protection and other services

| LifeLock protection features | LifeLock Standard | LifeLock Advantage | LifeLock Ultimate Plus |
|---|---|---|---|
| LifeLock Identity Alert® System | ✓ | ✓ | ✓ |
| Lost wallet protection | ✓ | ✓ | ✓ |
| Address change verification | ✓ | ✓ | ✓ |
| Reduced pre-approved credit card offers | ✓ | ✓ | ✓ |
| Black market website surveillance | ✓ | ✓ | ✓ |
| Live member support 24/7/365 | ✓ | ✓ | ✓ |
| $1 million total service guarantee | ✓ | ✓ | ✓ |
| Fictitious identity monitoring | | ✓ | ✓ |
| Court records scanning | | ✓ | ✓ |
| Data breach notifications | | ✓ | ✓ |
| Online annual credit reports and scores | | 1 credit bureau | 3 reports |
| Credit card, checking and savings account activity alerts | | ✓ | ✓ |
| Investment account activity alerts | | | ✓ |
| Checking and savings account application alerts | | | ✓ |
| Bank account takeover alerts | | | ✓ |
| Credit inquiry activity | | | ✓ |
| File-sharing network searches | | | ✓ |
| Sex offender registry reports | | | ✓ |
| Monthly credit score tracking | | | ✓ |

# Firewalls and Backup Memory

Corporations are now routinely using not only antivirus programs but firewalls as well. Most individuals rely on only antivirus protection, but corporations and individuals with home businesses feel this additional level of protection is useful. However, the protection as provided by personal firewalls can sometimes prove frustrating because you can be blocked from accessing sites like drop boxes that you actually wish to access.

A firewall is a coded protective barrier that separates a company's inside corporate intranet or your home-based "intranet" from the outside Internet or even your own smartphone in its most basic form. This typically allows effective and protected communications within a corporate environment, a home network, or just your desktop computer, tablet or smart phone. The idea is to create a barrier that "walls" you off from the Internet so that you have a filtered screen that keeps out harmful external communications.

A protected intranet, even with a sophisticated corporate firewall, is still no guarantee of absolute protection. If a telecommuting employee is at home, logs in, and then connects to an unprotected wireless LAN, then security is breached. Anyone with a scanner can eavesdrop via this open and unprotected access to the corporate network. Firewalls can also be penetrated by expert "crackers."

The following analysis from Brighthub indicates both the pros and cons of personal firewall protection [6].

## Common Features That Personal Firewall Can Offer

- It can help protect the user from unwanted incoming connection attempts—especially by spyware and keylogging software.
- A firewall generally allows the user to control which programs can and cannot access the local network and/or Internet. It typically provides the user with information about an application that makes a connection attempt.
- A firewall also can block and alert the user about outgoing connection attempts from within a local area network. (This is the case if the firewall is actually serving multiple computers rather than a single computer.)
- A firewall serves to hide a computer from port scans. This is accomplished by not responding to unsolicited network traffic.

The main function of the firewall is to monitor and regulate all incoming and outgoing Internet users and their traffic. It thus should prevent all unwanted network traffic from being able to affect (or infect) locally installed applications.

## Limitations of Firewalls

- If the system has already been compromised by malware, spyware, or keylogging software, these programs can also potentially manipulate the firewall. This is because they are both running on the same system. It may be possible to bypass or even completely shut down software firewalls, depending on their design.
- If a firewall has initially been incorrectly configured, this will not be obvious to the user.
- Firewall may limit access from the Internet, but it may not absolutely protect your network from wireless LAN access or and other access to your systems.
- Firewalls and virtual private networks are not the only solution to secure private documents and e-mails. Encryption or other techniques can be used.

- The frequent alerts that can be generated by firewalls can over time make users less vigilant concerning actual malware attacks.
- Software firewalls that interface with the operating system or with other firewalls or security software at the kernel mode level to allow backdoor access could potentially cause instability and/or introduce security flaws.

Large corporations with many users, such as airlines, large retailers, or credit card companies, hire "white hat" hackers to try to penetrate their firewalls in order to better protect themselves and bolster the protected shield of their firewall. Not all computer adept hackers are bad guys! The U. S. Cyber Command and lots of other units in the United States and around the world are using their white hat skills to protect against viruses and other malware that attacks the Internet and governmental and corporate databases and networking services.

The pricing for a firewall depends on the amount of traffic in and out of a computer or a computer intranet within a corporation or enterprise network. Individuals with personal firewalls would thus get protection at a relatively low rate for a small amount traffic that is equivalent to the cost of an antivirus. Some vendors offer free antivirus software with a firewall, for instance.

Large global corporations with so-called enterprise networks will pay thousands of dollars for firewall protection. Such corporations or governmental agencies, of course, typically also pay for back-up memory to store and protect vital data in a highly secure site. Today corporations often use the Cloud for storage and secure processing, but individuals should be sure the Cloud that is accessed is completely secure and backed up and well protected against cyber-attacks.

A 2013 assessment by NSS Labs of Next Generation Firewall (NGFW) systems for enterprise networks concluded that these products continue to improve in terms of qualitative performance. The NSS ran a series of demanding tests that gave recommended status to eight out of the nine systems listed below in terms of performance, but gave an unexpectedly low rating to the Palo Alto system. It also reported that there is a still a significant range with regard to the cost of protection. The NSS Labs report found that the Next Generation Firewall protection rates (per 1 megabit/s data streams) ranged for a low of $18 per 1 mbps to a high of $125 per 1 mbps (Table 2.5).

If one might doubt why such a sophisticated firewall is essential for corporations and individuals in today's universally accessed cyber world, you need only to recall the Sony hacking incident. This malicious cyber-attack mounted by North Korean sponsored hackers almost brought this media giant to its knees. It did not put Sony out of business, but the losses were in the millions of dollars and the exposed confidential messages led to enormous embarrassments and even suits against Sony by its clients and superstars.

Table 2.5  Assessments of firewall effectiveness

| NSS next generation firewall comparative analysis reports | |
|---|---|
| Next Generation Fire Wall | Rating results |
| Check Point 12600 | Above 90 % and recommended |
| Dell SonicWALL SuperMassive E10800 | Above 90 % and recommended |
| Fortinet FortiGate 3600C | Above 90 % and recommended |
| Juniper SRX 3600 | Above 90 % and recommended |
| Palo Alto PA-5020 | Not recommended |
| Sourcefire 8250 | Above 90 % and recommended |
| Sourcefire 8290 | Above 90 % and recommended |
| Stonesoft 3202 | Above 90 % and recommended |
| WatchGuard XTM 2050 | Above 90 % and recommended |

Go to NSS Lab site to find out how to get the detailed NSS Labs assessment: https://www.nsslabs.com/next-generation-firewall-reports

# Insurance Offerings

One can go beyond obtaining an antivirus, identity theft protection, or firewall system in order to isolate your network from intrusions. In this case you are seeking a comprehensive overall set of defense systems against all conceivable types of computer attacks. This can include such aspects as protection against cyber-bullying, junk mail, or loss of your purse or wallet.

In short there is a comprehensive approach to network security, identity theft, phishing, pharming, and any other possible form of cyber-attack on you, your family, or your assets—comprehensive data protection services in the form of an insurance policy. The MetLife Defender, for instance, provides this comparative analysis of their service versus a "typical" identity theft protection service. MetLife is, of course, just one such provider of this comprehensive insurance coverage. Again we are not necessarily endorsing Met Life. You should go to independent consumer advisor sites for help in making your decision. Again these sites include [7] (Table 2.6).

The insurance policy approach has its advantages in being comprehensive and providing cash compensation if something goes wrong, but it can also be expensive. Many such financial data production plans are available through employers, however, and thus at substantial discounts. A number of the antivirus, firewall, or malware protective services do offer up to $1 million in protective compensation against intrusions. If you are an individual the minimum protection that we recommend is an antivirus with a reasonably good rating and one that can allow you protection on all of your devices that connect to the Internet. Unfortunately, as we enter the age of the Internet of Things, you may have a number of appliances or devices in

Table 2.6   Comparison of comprehensive insurance offering from MetLife

| Feature/benefit | MetLife defender | Typical ID theft protection services | Typical financial institutions | Typical credit monitoring services |
|---|---|---|---|---|
| Financial data protection comparison | | | | |
| Patented multi-account financial protection | Y | N | N | N |
| Credit monitoring | Y | Y | N | Y |
| Lost wallet protection | Y | Y | N | N |
| Identity theft and personal data protection comparison | | | | |
| 25 data point personal identity monitoring and security | Y | N | N | N |
| Patented deep internet protection | Y | N | N | N |
| Privacy protection | Y | Y | N | N |
| Junk mail removal | Y | Y | N | N |
| Health data monitoring comparison | | | | |
| Medical insurance fraud protection | Y | N | N | N |
| Health and medical data protection | Y | N | N | N |
| Online child safety comparison | | | | |
| Child identity theft protection | Y | Y | N | N |
| Cyber predator detection and notification | Y | N | N | N |
| Cyberbullying applications and scanning | Y | N | N | N |
| Additional features comparison | | | | |
| 24/7 service & support | Y | Y | Y | N |
| Proactive alerts | Y | Y | N | N |
| Theft and fraud resolution service | Y | Y | N | N |
| Service guarantee | Y | Y | Y | N |

your home or car or truck that link to the Internet without you even being aware of these connections. In this case we will likely need legislation to force manufacturers to provide antivirus protection so that netbots cannot be used against you or others.

## What to Do

There are many practical steps that you can take to protect yourself. The minimum is a basic antivirus program. The antivirus that is often rated the most highly in terms of performance and cost is McAfee by Intel Security. Avast and Avira are the most widely used of the free antivirus offerings, but the *PC Magazine* website reviews all such offerings. From there one can consider obtaining a firewall and identity theft coverage. An identity theft protection plan such as the AARP offering in collaboration with Equifax is available to AARP member at just over $100 per year, but there are many packages that are available for a similar price that are likely comparable in coverage areas.

There are also minimum-level protection alert systems against identity theft or credit alerts.

Some companies such as ID Safe or AllClear ID offer free, bare-bones versions of identity theft protection services that monitor the Internet for hints that your personal particulars are for sale. These services would normally include monitoring of three credit cards to determine if the numbers are being sold on the black market plus a free annual credit report and score.

IDSafe or AllClear by Debix can also help you close accounts that have been attacked, place fraud alerts at the big credit bureaus and assist with the financial damage to help you restore your credit if your identity is indeed stolen. All of these services, just as in the case of Avast.com or Avira.com that provide "free antivirus," seek to upgrade you for a paid version that would typically cost $15 a month [8].

As noted at the outset of this chapter a full range of protection against computer viruses, a firewall, and up to $1 million in identity theft protection is available for substantially less than a year's subscription to the *Washington Post, The New York Times*, the *Chicago Tribune* or the *LA Times*.

At the top of level of personal cyber-protection coverage is a comprehensive "insurance policy" like the MetLife Defender. Corporations in today's world must recognize that their databases, key intellectual information, billing systems, etc., are their most valuable resources. Protection of these resources via all forms of cyber-defenses, including high quality encryption, is necessary in today's world. This book is focused on the individual. But if you own your company or work for a corporation, you should make sure that you have a full range of cyber-defenses in place as well. In the chapter on the future we will examine in more detail new ways to protect government databases and companies that operate vital infrastructure. These cyber-protections are now key to families, businesses, and national security.

## And Now How to Protect Your Family Against Cyberbullies, Pornography, Online Hate Messages, and Other Cyber-Related Maladies

First, there is the straight talk message that protecting you and your family against unwanted content over the Internet is difficult, and law enforcement officials are often slow to respond to help and in addition are limited in the laws that they have at hand to enforce. Further, the most heinous and hurtful trolls, spinner of false talks, purveyors of hate crimes, and distributors of pornography hide behind anonymous web names and unidentifiable web domain names to avoid criminal prosecution. Blocking purveyors of hate messages and pornographers and legally prosecuting them or shutting them is, unfortunately, hard to do.

The reasons why cyberbullying, hate messaging and pornographic distribution is so hard to get rid of are anonymous websites, web postings, and aliases. Thus the attackers and trolls are often "ghosts," and, as you can see from the list below, it is hard to locate these spectral beings that only exist online and accordingly can hide their real identity.

1. Many of the most offensive websites and sources of hate crime messages and pornography operate from offshore sites outside of the United States, Europe, or other countries that have enacted laws to prosecute cyber-criminals and techno-terrorists.
2. Targeted individuals are often frightened by such attacks and are reluctant to bring the offensive language, false profile, or Photoshopped™ image to the attention of parents, school officials, or the police. A young girl who has had their face pasted on the provocative nude image of a Playboy Playmate may be too embarrassed to report this to their parent or school advisor.
3. There are a lack of effective laws and ordinances that allow criminal prosecution of offenders. And court cases take time and money to pursue. While a legal case is going through a prosecuting attorney's office and through a trial, the offending material could well remain online.
4. Many cyber-attacks are protected by freedom of speech and the right to express oneself openly in democratic societies.

And the problem is not a minor issue with people affected by malicious cyberbullying or hate crime attacks. The *2013 Youth Risk Behavior Surveillance* Survey finds that 15 % of high school students (Grades 9–12) were electronically bullied in the past year. It is noteworthy that there was a wide variation in gender of those actually bullied, namely 21 % of girls but only 8.5 % of boys [9].

The *Journal of the American Medical Association* found 50 % of children have been asked to send revealing or even nude pictures of themselves over the Internet, a practice now widely known as sexting. In a world filled with X-rated videos, adult video channels, and cable television channels in many homes with pornographic programming, it is hard for young people to understand what is wrong with sharing racy photos or even nude images.

Even if there is no such programming in your home or V-chip restrictions on adult programming, this type of programming is likely available at a friend's home instead [10].

According to interviews conducted by Pew Studies and media-sponsored surveys, only 50 % of parents talk with kids about avoiding these encounters. Teens in particular try to cope on their own. This can result in retaliatory behavior that can put youths even more at risk, or can let a problem grow out of control. By the time adults, scholar advisors, or even police become involved in Internet-based controversies hundreds if not thousands of others have become targets of cyber-attacks that may involve false profiles, allegations of wrongdoing or other hurtful information being spread via text, social media, or website postings that could ultimately hurt your offspring's chances when it comes to college admissions or job interviews.

And it should be noted that the problems of the cyber-world continue to expand as we become more and more dependent on our "intellectual prosthetics." Today our "smart" computers, laptops, and interactive smart phones can provide almost any requested information on instant demand. With hundreds of apps now available to us, there can be a loss of spelling, math, or reasoning skills. Most young people today who rely on texting to communicate do not know how to write cursive, spell words, and are starting to lose basic skills in mathematics. In a world where "spell check" and "smart phone" and other intellectual prosthetics hold sway there are concerns that this loss of skills that is sometimes called "de-skilling" will exacerbate the problem of competing in a global information services market. Copyediting, inventory control, back office accounting, and software development can now be done (if poorly) by anyone in the world with a communications link and computer (Fig. 2.1).

## What to Do About Protecting Your Family's Computers from Cyber Criminals, Stalkers and Bullies

The first thing to do is to recognize that these are real problems that are quite common. You need to assume that various types of cyber difficulties will actually arise for your kids, your spouse and/or elderly parents or relatives. You need to talk to them and warn them of potential problems that can arise and are now common they are now around the world. As noted in the above studies by many reputable research organizations cyberbullying

**Fig. 2.1** Problem of de-skilling is yet another worry that comes with the computer age (Graphic provided by the author J. Pelton from his book *Future View.*)

via social media, texting, and other electronic means is a growing problem, and because of the Internet more people can be exposed quickly and the harm can be enormous. It has alarmingly led to a rash of suicides in countries around the globe.

Fortunately there are videos and online materials that can help spell out the problems, potential solutions, and websites and services designed to assist with these various cyber problems. The following table indicates recommended videos, websites, and services designed to cope with cyber problems associated with teens and "tweens" (those not quite teens, or in U. S. parlance, "middle schoolers"). These sites and services can also assist parents in coping with cybersecurity and especially cyberbullying problems involving their children. There are a growing number of services that address not only cyberbullying but also monitoring smart phone usage and restricting viewing patterns for kids (Table 2.7).

Table 2.7  Available websites, services, and videos that you and your family might use to find advice about cybersecurity and cyberbullying

Federal Bureau of Investigation (FBI) Video on Cybersecurity
Go to: https://sos.fbi.gov
This website offers fun, age appropriate lessons (3rd grade up) about cyber safety. The exercise included is based on an interactive island exploration game. It is an educational game and definitely more fun and interesting than listening to another safety lecture from an adult.

Sponsored by Trend Micro "Cybercrime exposed: how to spot a phishing scam"
A 2:20 min. video on youtube.com
Go to: www.youtube.com/watch?v=pXp2RvA0SBU

Sponsored by AVIRA SocialShield Avira.com
"How to Protect Your Kids on Social Media,"
A 3:27 min video for parents.
(Note Avira.com is one of the providers of a "free" antivirus service along with Advast)
Go to: www.youtube.com/watch?v=IRlAMMaNr8U

Federal Trade Commission (FTC) "Stand up to cyberbullying" (1:20 min)
Go to: www.youtube.com/watch?v=lN2fuKPDzHA

BRIM website
Go to: https://stopbullying.gov
This site provides useful tips about how to cope with various attempts at cyberbullying and strategies that can help protect children. This you should view with your children and answer their questions.

HIBSTER website
Go to: http://hibreporting.com
This website is an active anti-bullying program that specializes in incident prevention, management, and reporting.

STOPIT
Go to: http://stopitcyberbully.com
How to report, track and effectively deter cyberbullying in your school and neighborhood. When a child on "StopIT" sees something online that's not okay, whether directed at that person or a friend or peer he or she can immediately use the STOPIt app to let a trusted adult know right away what the cyberbullying message is about. Those signed up for "StopIt" can send messages and/or screenshots directly to the trusted adult(s) such as a parent, aunt or uncle, grandparent, or school counselor. This can even be sent anonymously to alert a responsible adult. Details on enrollment are pending shortly.

App Certain
Go to: https://www.appcertain.com
Cost: Free
App Certain will e-mail parents when their child downloads a new app, and will provide an analysis about that app like if the app has expensive in-app purchases or accesses your contact list. Parents can also utilize a "curfew mode" that gives the remote access the ability to turn off their children's access to their apps and games at a certain time.

Mobile Watchdog
Go to: https://www.mymobilewatchdog.com/default.shtml
Cost: Free for 14 day trial and then $4.95 per month, with cancelable service at any time.
Allows users to monitor all cell phone activity on Android and I Phone devices—text messaging, application use, and browsing use. You can program in special words and be alerted if these words appear in a text or e-mail. This app will send you an e-mail reporting on a child's mobile phone activity. It does not have a "stealth operating mode" for listening in on cell phones.

# What to Do About Pornographic Sites and Sexting

There are many services that can introduce monitors for smart cell phones and use browsers that are geared not to access X-rated or pornography sites. Some of these are noted in the table below. The problem is that even if V-chips are used to block out pornographic cable television channels and some of the services such as K-9 Browser, young and impressionable youths can be exposed to nudity, sexual videos, and other undesirable behavior through the smart phones of friends or computers in other houses. The survey conducted by doctors for the *Journal of the American Medical Association (JAMA)* about sexting and teen sexual behavior is indicative of the contemporary sexual environment is in the United States, which is probably not greatly different from other OECD countries. This particular study found that nearly 25 % of 16-and 17-year-olds had experienced sexting on the sending or receiving end. Further its findings indicated that those who had engaged in this practice were twice as likely to have engaged in sexual intercourse (i.e., 77 % versus 40 % of those that had not sexted) [10].

This is simply to say that monitoring and controls can only go so far to protect children—particularly when they reach a certain age. Once children reach perhaps 13 or 14 years old such controls really can easily be avoided by enterprising youngsters. If they feel they are being overprotected or their viewing habits unduly restricted this might well lead to rebellious behavior. Good and trusted communications between parents and offspring and a good academic environment with trusted teachers are key to healthy development. It is really a matter of instilled values, and trustworthy friends and siblings, rather than imposed limits once a tween approaches teenagehood (Table 2.8).

# Protective Strategies for the Over-50 Crowd

One can surf the website of the American Association for Retired Persons (AARP) and find a number of very useful tips and strategies on digital privacy, cybersecurity, backing up vital information stored on computer hard drives, and protecting against phishing, pharming, ransomware and identity theft. These can be summarized in the following list.

Table 2.8 Services to restrict access to pornographic sites and monitor smart phone usage

| |
|---|
| K-9 Browser |
| Cost: Free |
| This provides a specially designed and highly rated browser that children can use instead of an Internet browser that goes to all possible sites. This browser with this special app will block adult content. It's available for the iPad, iPhone, iPod, Android, and a desktop computer. The difficulty is to ensure that children consistently use this browser and not Yahoo, Google, or another browser with no screening capability. |
| Net Nanny |
| Cost: See website for details |
| This service can be installed on I-Phones and Android phones for various annual fees ranging from a low of about $5 up to $20 for Net Nanny Social. |
| Net Nanny has mobile monitoring services for Android and Apple that will help block adult content. It also offers Net Nanny Social, which installs special software to screen for cyberbullying or unsafe activity. If anything unsafe is detected, parents receive an alert. Parents can also log in and see all social media activity in a Cloud-based dashboard on any device. |

1. Protect the data, photographs and key documents that you have stored on your computer's hard drive. As noted on the AARP site, your computer's hard drive, because it is a mechanical device, will ultimately fail. You can key keep a lot of key information on a thumb drive (also known as a flash drive), but this can't store everything. There are lots of high density hard drives (portable storages of several hundred gigabytes and desktop units of several terabytes). These run from $50 to $200 (e. g., Seagate, Western Digital, etc.). Or there automatic backup services that you can sign up for a small annual fee [11].

2. It is particularly important to protect against identity theft and to avoid the problems that can come with the theft of credit card numbers and codes, Social Security numbers, or Medicare and medical records. AARP articles highlight that 16 million people in the United States alone fall victim to identity theft, and that those over 50 are often sought out as prey for cyber criminals. AARP research notes that this is not always cyber theft, and highlights that only one in five people shred documents that could be used to steal their identity. It is interesting that articles on the AARP site explain that there are "free" services such as ID Safe and AllClear that provide a "bare bones" service, even though AARP itself provides a discounted rate for identity theft in its partnership with "Trusted ID." This is for a more robust service that has been given top ratings by independent reviewers, as noted in the table above [12].

3. It is important to install antivirus and a basic firewall on your computers and electronic devices, including your cell phone. Also make sure your wireless router is password protected, and don't go to sites that your antivirus software warns you not to go to [13].

4. Take special care to protect your "smart phone." One in three robberies net a cell phone. "Apple Picking" (or theft of I-Phones) is by itself a huge industry. A particularly useful AARP article outlines 12 ways to protect your smart phone [14]. Also you should be careful about how you sign up for and protect the insta-payment systems that are installed on your cell phone in terms of tap and go or swipe and go systems. (Note: These issues of cell phone security and touch and go systems are discussed in further detail in Chap. 5.)

5. Use social media wisely. Social media is a great way to interact with family and close friends, but be careful about posting too much personal information on your website that might be used by cyber criminals. This information might not only be used to try to get money or assets from you and your family, but it might also be used in phishing or pharming schemes to make a request to friends or family for money. If a cyber-criminal is armed with enough personal knowledge of you and your habits that person might exploit unsuspecting close associates of yours [15].

## Conclusions

If there is a vital message in this entire book, it is the following words of advice and counsel. The Internet, the world wide web, hyperlinks, social media, smart phones, texting, mobile apps are wonderful new tools of our times. These tools let us stay in touch with family, friends, and loved ones, buy what we need on the go, and live a more convenient and free-wheeling life. Electronic media and communications devices keep us educated, informed and up-to-date in ways that were never before possible in the entirety of human history. The amount of information that we can obtain, store and access is greater than ever before, and the cost of storage, processing, and imaging keeps falling as new computer and communications technologies move forward.

There is, however, a cost associated with the use of these electronic tools, one that we must pay to insure that we maintain our privacy, protect ourselves from identity theft, and increase the security of our family and loved ones. We must invest in protective systems and cybersecurity tools in order to ward off the various malware and viruses and worms that cyber-criminals might employ when they seek to infect our computers, laptops, tablets, smart phones, and other electronic gizmos.

In a word, you really need to obtain help from various vendors of protective services, consistently use and protect well-crafted passwords, and back up important and sensitive data against computer failure as well as to protect against an attack by a computer virus. For much less than what one might pay each year for a newspaper subscription, you can get a range of protective services as well as systematic help with other cyber-related problems such as cyberbullying, children accessing pornographic sites, etc. This is the trade-off. On one hand you can access a great deal of information and images and advice for free. In order to ensure that some of that information is not a harmful attack on you and your computer or smart phone you must wisely follow the advice provided to you in this chapter and sign up for cybersecurity services.

If your budget is particularly tight then at least avail yourself of free services such as those provided by Avast, Avira, IDSafe and AllClear. There are lots of helpful websites out there to aid you in making good choices. The main thing is not to delay putting protective cybersecurity systems in place. Also, you should follow other good privacy and security practices such as protecting your mail, shredding your financial and other personal records, and making sure your various important financial, medical, and personal passwords are backed up and well protected.

We can't cover every issue in one short book, but we hope this chapter in particular has started you on the way. Remember, there are lots of good and useful websites out there as well as public librarians and other public servants that can provide useful and timely advice.

# References

1. Norton Internet Security www.Symantec-Norton.com/Security.
2. "Top Ten Antivirus Sites" Matchtop.com. http://antivirus.thetop10sites.com/?matchtype=b&keyword=norton%20antivirus%20vs%20mcafee&adposition=1t1&creative=60402236670&aceid=&gclid=CMm88LWgh8MCFQhk7AodN3AASg#gsc.tab=0.
3. 2012 Identity Fraud Survey Report. Javelin Strategy & Research. February 2012.
4. "Top Ten Identity Theft Protection Sites." www.top10identitytheftprotection.com.
5. Life Lock Individual Plans: Identity Theft Protection, http://www.lifelock.com/services/.
6. Daniel Brecht and edited by: Bill Bunter, "Pros and Cons of Computer Security Systems" http://www.brighthub.com/computing/smb-security/articles/32411.aspx also see. Finn Orfano and edited by Bill Bunte, "Firewall Basics Part III: The Pros and Cons of Firewall Methods." http://www.brighthub.com/computing/smb-security/articles/8293.aspx.

7. MetLife Defender. https://www.metlife.com/individual/employee-benefits/metlife-defender/index.html.

8. "She Stole My Life," *AARP Magazine,* Nov. 2014.

9. "2013 Youth Risk Behavior Surveillance Survey, Centers for Disease Control and Prevention," June 13, 2014.

10. Jeff R. Temple, Ph.D., Jonathan A. Paul, Ph.D., Patricia van den Berg, P.hD., Vi Donna Le, B.S.; Amy McElhany, B.A.; Brian W. Temple, M.D. "Teen Sexting and Its Association With Sexual Behaviors," *Journal of the American Medical Association,* September 2012, Vol. 166, No. 9 http://archpedi.jamanetwork.com/article.aspx?articleid=1212181.

11. Gary M. Kaye, "Sooner or Later Your Hard Drive Will Fail: Computer backup is like digital fire insurance" AARP, June 23, 2011. http://www.aarp.org/technology/privacy-security/info-06-2011/back-up-computer-files.2.html.

12. Gretchen Anderson "Identity Theft: Who's At Risk?" AARP Research, September 2014. http://www.aarp.org/money/scams-fraud/info-2014/identity-theft-incidence-risk-behaviors.html.

13. Kim Loop, "Prevention, Not Just Awareness, Key to Cyber Security." AARP Texas October 11, 2012. http://states.aarp.org/prevention-awareness-cyber-security/.

14. Sid Kirchheimer, "Scam Alert: How to Cyberproof Your Phone – 12 ways to protect yourself and your device," AARP Bulletin, May 2014 http://www.aarp.org/home-family/personal-technology/info-2014/cyberproof-stolen-phone-kirchheimer.html.

15. Technology, Tablets, and Social Media. http://states.aarp.org/prevention-awareness-cyber-security/.

# 3

# Is Anyone Looking Out for You? Your Government? Businesses Where You Trade? Your Neighbor? or Just Yourself?

The question "Is anyone looking out for you?" is tantalizingly ambiguous. It can mean is someone looking out for your interests and trying to protect them? Or it could mean something else entirely.

It might mean there is big government looking to see if you have paid all your taxes or committed some crime for which you should be prosecuted. Or it could mean that there is a sophisticated industry team of computer analysts and hucksters who are using massive amounts of big data to figure if they could sell you something else? Or it could be a cyber-criminal looking into your affairs to see how to drain your bank account, glom onto your credit card information or even steal your identity for a nefarious purpose.

From the old days of "The Twilight Zone" TV show there was the classic story of the advanced alien race that came to Earth and seemed to helping humanity in many ways and flying happy human tourists off to their planet for extended stays. Then a book from the aliens is discovered, and the linguists and computer decoders first come up with the book's reassuring title, which is "To Serve Mankind." At the end of the "Twilight Zone" story the lead scientist is getting on board the starship for his own journey to the planet of the aliens when his aide rushes up to shout to his departing boss. "Don't go. It's a cookbook."

We live in a world of onrushing technology. It is a world where there is more and more automation, more storage of massive amounts of data about individuals and technology, and an ever accelerating rate of technological innovation. The inevitable question arises as to whether new invention and more "efficiency" in society actually translates to a better—or worse—life for the man or woman in the street. There are lots of dystopian science fiction stories and even general literature out there that suggested advances in medical technology, electronic communications, and computer technology

© Springer International Publishing Switzerland 2015
J.N. Pelton, I.B. Singh, *Digital Defense*, DOI 10.1007/978-3-319-19953-5_3

will lead us into a more sinister world of terrorism, cybercrime, and perhaps oppressive governments. *Brave New World,* by Aldous Huxley, *1984* by George Orwell, *Fahrenheit 451* by Ray Bradbury, *Player Piano* by Kurt Vounnegut, Jr. *I, Robot* by Isaac Asimov, or *2001: A Space Odyssey* by Arthur C. Clarke all, in their various ways, forewarn us of a future where technology advancement and the desire to create a better world for human existence will clash. Even in the last few years writers such as Dean Koontz, Michael Crichton, Robin Cook, and Neal Stephenson have written some very cautionary tales about a future society where technology—abused by human leaders and immoral scientists and engineers—can lead us down dark pathways that would best not be taken.

Fortunately we are still a long way away from such dark futures in democratic societies. Yet only careful vigilance will keep societies free, democratic, and aimed in the direction of providing liberty and justice for all. To protect the future for our offspring, it is wise to pursue an information culture based on the following three principles:

- Cybersecurity acts to protect against cyber criminals and cyber-terrorists.
- There is freedom of access to information without censorship, undue legal, political, social or cultural limitations.
- Cyber privacy avoids excessive governmental or industrial snooping to the maximum extent possible within a cyber-world that features more and more electronic broadband communications and computer networking systems prone to eavesdropping and spying.

There is, of course, a problem here. These three principles can come into conflict with one another and often do so. In theory, in a democratic society, governments protect their citizenry against external threats, defend them against abuses that jeopardize their liberties and freedom, defend them against unwarranted surveillance, and uses censorship very sparingly. When the protective function against external threats, for instance, requires wire-tapping and cyber-spying, these principles come into conflict. The right thing to do and which principle to defend becomes murky. In the heat of a terrorist threat the public may be willing to give up some privacy and some freedoms, but in the longer term they often regret giving too much snooping power to a government.

Likewise, industries and commercial enterprises are supposed to look out for the interests of their customers. Honest treatment of customers helps to bolster brand loyalty. In the high-tech world of the Internet networking and online sales, information about customer preferences, and desire to know if a particular line of commercial products suit customer needs, companies are tempted to compile information on consumers likes and dislikes and create databases that look a great deal like corporate spying. In the world of "big data," market data accumulation can become an intrusion into consumer privacy. Further, if corporate data is compromised and then obtained by criminal interests, the problems can be compounded.

From yet another perspective, if corporate cybersecurity is lax, vital corporate infrastructure, such as privately controlled nuclear power plants, railway switching systems, or telecommunications signaling and switching centers are easily subject to cyber-attacks, and the dangers escalate. There is indeed mounting evidence that many corporately managed rail, telecommunications, power, airline, and pipeline systems do not have adequate cybersecurity protections and are thus exposing the general public to grave and undue risks.

The long and short of it is that both governments and industry expose us to a number of risks each day. As our society becomes more and more dependent on cybernetic electronic controls, "smart" networking, and automated financial transfers and payment systems our vulnerability grows. Our money, our Social Security payment systems, our stock holdings, or power, communications and transportation systems, and our privacy all can be compromised—or much worse—endangered or destroyed. Even experts who reviewed this book were reluctant to concede the incredible damage that techno-terrorists could cause to nuclear power plants, the air traffic control system, or water treatment plans.

Governmental agencies and the courts from one perspective and corporations on the other are torn about how to know what is the right thing to do. In a crisis one course of action may seem right, but after the crisis is past too much surveillance and use of cyber tools to spy on the populace in a systematic way can become oppressive. Honoring and obeying these three key principles is one of the great challenges in today's democratic societies. As we transition to a more automated world controlled by networks and artificial intelligence the problem will not diminish. It will only become worse.

This is the particularly tricky challenge of our times. The world of the future with the Internet of Things, fifth-generation smart phones, and "smart infrastructure" controlled by knobots (or knowledgeable robots) puts us at risks in many ways. Cyber criminals, cyber-terrorists, breakdowns due to natural disasters for which we are not prepared, and lack of privacy are all significant risks shared by developed economies all over the world. They are real problems that must be addressed with greater forethought, wisdom and caution than we have up to this point in history. In the rest of this chapter we will examine these risks in three sections: (1) risks that come from governmental cyber policies and programs; (2) risks that come from industry cyber policies and programs; and (3) risk mitigation and cyber defenses that individuals and families should take on themselves.

## Governmental Issues: The "Snowden Factor"

In 1978 the U. S. Congress created what is called the U. S. Foreign Intelligence Surveillance court (or more simply the FISA court). This is a federal court established and authorized under the Foreign Intelligence Surveillance Act of 1978 (FISA) to approved wiretaps or online monitoring of terrorists or those suspected of being foreign intelligence agents inside the United States. As an aftermath to the 9/11 attack on the Pentagon and World Trade Center, the U. S. Congress approved the Patriot Act. This act further empowered the U. S. intelligence community to monitor the activities of terrorists and would-be terrorists and gave an even more pivotal role to this unusual court that always heard the prosecutor's case but never heard from the defendants and their counter case.

This special FISA court that largely acts on the request of the Federal Bureau of Investigation (FBI) and the National Security Agency (NSA) is very responsive to requests for electronic surveillance and purportedly approves some 90–95 % or more of all requests for surveillance. In response to the criticism FISA court presiding judge Reggie Walton and California senator Dianne Feinstein have noted that nearly 25 % of requests are modified and often narrowed in scope [1].

In 2013, a top-secret order issued by the FISA court was leaked to the media by Edward J. Snowden, a computer expert and CIA administrator, with significant press coverage about its activities and special clandestine rulings. The Snowden materials revealed a FISA court order that apparently required Verizon, through a subsidiary unit, to provide on a daily basis a detailed record of all calls—including those for domestic calls to the National Security Agency. The speculation is that if this were the case for Verizon it was also true for other U. S. telecommunications carriers.

In July 2013 *The New York Times* disclosed further disturbing information about the FISA court. It published accounts from "whistleblowers" whose identities were protected. These informed sources discussed what amounted to what they claimed to be a "secret law" whereby it was deemed to be legal to compile and hold a vast collections of data on all Americans

**Fig. 3.1** The National Security Agency and the FISA court are viewed by some as the 'Big Brother' of the twenty-first (Graphics from author Joseph N. Pelton's book *Future View.*)

(even those not connected in any way to foreign enemies) that had been and is still been collected by the NSA. The ruling was that the collection and holding of such secret data did not violate the explicit "warrant requirements" of the Fourth Amendment to the U. S. Constitution and its "special needs" provision. *The New York Times* article reported that anyone *suspected* [emphasis added] of being involved in nuclear proliferation, espionage, or cyber-attacks may be considered a legitimate target for warrantless surveillance. The report thus suggested that the FICA court, in its operations, was functioning like a *sub rosa* U. S. Supreme Court in providing final determinations as to what is constitutional or not [2] (Fig. 3.1).

In the case of the FISA court, the right to privacy and the right to have a warrant served to alert someone that they are being placed under surveillance or charged with some sort of a crime is today being trumped in the U. S. legal system by the greater importance of the anti-terrorism function. Those suspected terrorist activities and serious cyber-crimes—especially those involving an attack against the United States and its citizens—are today subjected to warrantless surveillance and no notification.

Perhaps the most sensational of all, from Snowden's revelations, was that there was widespread surveillance by the U. S.'s NSA of millions of foreign nationals. French President Francois Hollande, in response to the Snowden-released documents, complained in a 2-week period between the end of 2012 and early 2013 that the NSA had collected some 70 million electronic messages involving French citizens and diplomats. The most heated reactions came from Germany when it was revealed that the NSA was even monitoring the private cell phone of German chancellor Angela Merkel and other heads of state [3].

And this situation is in no way unique to the United States. Many democratic societies around the world have similar processes at work to combat terrorism, track cyber-criminal activities, and use telecommunications systems and the Internet to monitor the communications and messaging. The point is that the revelations in NSA documents released by Edward Snowden indicate that many others are being monitored as well.

The most recent heated discussions of this practice surfaced in the United Kingdom in early 2015. The Communications Data Bill, that is known unofficially in the U.K. press as the "Snooper Charter," was supposedly dead in the 2015 sessions due to Parliamentary objections. Yet, the key provisions to allow surveillance and store so-called metadata related to calling patterns for all citizens, are still contained in the amendments to the Counter-Terrorism and Security Bill (CTASB), which apparently will actually pass. The amendments to this CTASB bill broadly describe a data retention regime that allows the storage of data on who calls whom. And this new bill only adds to the powers already given under the existing U.K. Data Retention and Investigatory Powers (DRIP) act. This act was passed as "emergency" legislation that allowed the British government to engage in "data retention" related to U.K. citizens after the EU banned its earlier Europe-wide mandate. Although protests to U. S. electronic surveillance activities have been heard far and wide, due to Snowden's release of NSA documents, it seems that there are parallel practices in many European countries [4].

The European system currently requires the cooperation of some 200 different intelligence, policy, and defense units. Such complex cooperative arrangements have been frequently been critiqued as being ineffective. Critics argue that the realistic chances of getting secretive national security agencies that are normally unwilling to share data, and then demanding that they share highly sensitive information, amounted to "wishful thinking," especially if there is no legislative mandate behind such an initiative.

The revelations by the American whistleblower Edward Snowden about agency surveillance has, of course, raised other questions. Critics of government "snooping" have questioned whether the law enforcement agencies taking part in the drill should be involved in safeguarding online security, in light of documented online spying by western governments.

"The main concern is national governments' reluctance to cooperate," said Professor Bart Preneel, an information security expert from the Catholic University of Leuven in Belgium. "You can carry out all of the exercises you want, but cybersecurity really comes down to your ability to monitor, and for that, national agencies need to speak to each other all the time," Preneel said. The European Union's Agency for Network and Information Security (ENISA) is based in Crete, but it does not have a legislative mandate to force national agencies to share information. This unit's domain has been described as only to provide a body of expertise.

As with most aspects of policing and national security, the EU's 28 members have traditionally been reluctant to hand over powers to a central organization, even when—as in the case of online attacks—national borders are almost irrelevant [5]. In countries such as Russia, China, Cuba, and many other states the monitoring of citizens' telephones, cell phones and Internet communications is considered a normal way of life [4]. The point is that many people are assuming they have a level of privacy that simply does not exist.

In addition, there is today an ongoing struggle between Internet service providers and governments about what types of encryption should be used in electronic communications (Fig. 3.2).

A recent public forum featured NSA head Michael Rogers and Yahoo's chief security officer Alex Stamos. This very hostile encounter was characterized in *Slate* magazine in the following manner: "From the perspective of many in the American tech industry, the NSA's actions represent an "Advanced Persistent Threat" similar to the cyber threats posed by organized crime or Chinese intelligence, while also threatening their bottom line by undermining worldwide consumer trust in the security of "American companies' products" [6].

This conflict arises from NSA's persistence in covertly listening in on Yahoo, Google, and Explorer private data links and collecting vast amounts of data that users consider to be confidential. Nor is this the only privacy issue at hand. Apple's I-Phones and Android smart phones by Samsung and other suppliers can currently be "hacked" by the NSA. It has also been revealed that the codes needed to hack into SIMs (subscriber identity

**Fig. 3.2** Electronic or cell phone privacy vs. governmental surveillance: A losing battle? (Graphics from author Joseph N. Pelton's book *Future View*.)

modules) embedded in the "SIM firmware" have been obtained by the NSA. Alex Stamos of Yahoo asked why should his company or others "build back doors into their encrypted products to facilitate government surveillance, when all the technical experts say that cannot be done without opening users up to threats other than the government" [7].

The gigantic data center that the NSA has constructed in a remote part of Utah will only escalate the agency's capability to monitor communications on an almost gargantuan scale. In short the NSA continues to exploit security vulnerabilities in a range of U. S. tech companies' hardware and software products. In a clash where the outcome is unclear tech companies such as Google, Yahoo, Microsoft, Apple, and Samsung are starting to employ encryption to try to shut out the NSA's snooping. The huge $2 billion facility in Bluffdale, Utah, has the innocuous name of the Utah Data Center, but it will have the ability to store almost unlimited amounts of data. The purpose of the Utah Data Center has been described as "to intercept, decipher, analyze, and store vast swaths of the world's communications as they zap down from satellites and zip through the underground and undersea cables of international, foreign, and domestic networks." The heavily fortified $2 billion center came into operation in late 2013. Flowing through its servers and routers and stored in near-bottomless databases are all forms of communication, including the complete contents of private e-mails, cell phone calls, and Google searches, as well as all sorts of personal data trails—parking receipts, travel itineraries, bookstore purchases, etc.

In addition this center employs leading cryptologists to break through encoded messages. This awesome capability that NSA has established in remote Utah resembles the discredited "total information awareness" program envisioned by Admiral John Poindexter and DARPA (Defense Advanced Research Projects Agency) official during the presidency of George H. Bush. The effort was defunded by Congress in 2003 after widespread public complaints that this would constitute a massive invasion of Americans' privacy. Yet today a little over a decade later the Utah Data Center is at the heart of what seems to be total information awareness redux [8].

Just as we can see the government as a problem in the cyber world by invading citizens' privacy and encroaching on individual freedom, it is also a key source of protection against cyber-attacks. There has been a concerted effort to shore up the cyber security of corporations and industry that provide financial services and vital infrastructure for modern society. The U. S. government agency known as the National Institute of Standards and Technology (NIST) has developed a security framework for industry to follow to provide a higher level of protection against cyber-attacks that is detailed in Chap. 5. There has been an effort to enact legislation in the United States to enforce much higher levels of cybersecurity, and governments in other countries have parallel efforts. In the United States, the Cyber Intelligence Sharing and Protection Act (CISPA) has been up for consideration in the U. S. Congress in various forms since 2012. It has been strongly criticized by organizations such as the American Civil Liberties Union, the Electronic Frontier Foundation, and others that are concerned about the loss of privacy that comes with the pursuit of cyber-terrorists. It seems likely that some version of this act—after some four years of negotiation—will be passed. In December 2014 President Obama summed up the current situation as follows:

> In this interconnected, digital world, there are going to be opportunities for hackers to engage in cyber assaults both in the private sector and the public sector. Now, our first order of business is making sure that we do everything to harden sites and prevent those kinds of attacks from taking place…But even as we get better, the hackers are going to get better, too.

Some of them are going to be state actors; some of them are going to be non-state actors. It's…important for Congress to work with us and get an actual bill passed that allows for the kind of information-sharing we need. Because if we don't put in place the kind of architecture that can prevent these attacks from taking place, this is not just going to be affecting movies, this is going to be affecting our entire economy in ways that are extraordinarily significant [9].

Congress now appears to be on the verge of adopting a cyber-bill that would allow firms to share information with U. S. governmental agencies that involve "cyber-threat indicators" without penalty. These indicators would include "malicious software," "parts of an e-mail," or "Internet protocol addresses" that suggest a techno-terrorist threat. The pending legislation from the U. S. House of Representatives and the Senate does have some privacy protections but would allow a new real-time alert to be sent to the Department of Homeland Security (DHS). The DHS would be empowered under this pending legislation to alert the Department of Defense and the National Security Agency [10].

Virtually all governments are seeking a strengthened capability in their external and internal capacity to track and prosecute cyber criminals and especially to intercept cyber-terrorists. In the United States this includes the Federal Bureau of Investigation (FBI), the Central Intelligence Agency (CIA), the National Security Agency (NSA), the Department of Defense (DoD), plus the Defense Advanced Research Projects Agency (DARPA), the National Institute of Standards and Technology (NIST), and the Department of Homeland Security (DHS).

In Appendices B, C, and D of this book the current status of U. S. and international cybersecurity programs are provided in form of a review of these activities. These include materials related to the United States, including the Comprehensive National Cybersecurity Initiative (CNCI) as well as reports on the Japanese new Cyber Security Defense Unit and the Organization for Economic Cooperation and Development (OECD) guidelines for the Security of Information Systems and Networks.

# Corporate and Industrial Issues

In many ways the roles of industry and government in the cyber world are reversed. Industry defends itself vigorously against privacy snooping by governments. Organizations such as the Electronic Frontier Foundation and companies such as Yahoo, Google, and Microsoft have stood for encryption, Internet privacy, and resistance to inserting "back doors" for government spy organizations for years. They have endorsed and supported efforts like "PGP" (or pretty good privacy) encryption as a way to shield one's own personal communications [11].

However, when it comes to protecting the interests of customers and defending against cyber criminals and cyber-terrorists, industry has shown less focus than governments on this important mission. The various efforts by governments to make sure that industries provide the highest level of cybersecurity has not been well heeded. The United States and other governments, as discussed above, have made a concerted effort to make sure that companies that provide vital infrastructure and conduct financial and credit card transactions are particularly well shielded against cyber criminals and especially techno-terrorists.

As noted above measures such as the Foreign Intelligence Surveillance Act, the Patriot Act, the Critical Infrastructure Protection Act, and other pending measures in the U. S. Congress often find industry and government at odds as to what to protect and how vigilant the protection process should be and who should be in charge of the protection process. The same sort of divide can be seen in the U.K. with the Communications Data Act, the so-called "DRIP" act, and "Snooping Charter." Just about all of the members of the Organization for Economic Cooperation and Development (OECD), which includes most of the economically developed democracies, this dynamic tension remains in effect.

Currently the extension of the Patriots Act is pending in the U. S. Congress. There are a number of groups that are seeking reform of the massive collection of so-called "metadata" that is being collected by the NSA on a massive scale. The Reform Government Surveillance industry coalition that includes Google, Microsoft, Apple, and Twitter has sent a message to Congress and President Obama stating: "The status quo is untenable.... It is urgent that Congress move forward with reform." Section 215 of the Patriot Act, that purportedly allowed the secret metadata collection by the NSA, was allowed to expire without renewal [12]. The newly enacted U.S.A. Freedom Act now signed into law makes it clear that the U.S. government can no long collect massive amounts of metadata on phone calls and instead must get a warrant from the FISA Court to get call records from telecommunications carriers. This is one of the consequences that followed the Edward Snowden revelation about the extent of government snooping through elaborate computer record-keeping systems [13].

Today the general citizenry often tends to feel somewhat left out. Industry increasingly feels entitled to use social media, billing records, and a host of other online data to analyze and project consumer spending habits and inclinations. Most people automatically click on "Agree" to terms and conditions of use to access a corporate website without knowing which rights they have just forfeited. In some cases they have agreed that their e-mail address and other data about them can be sold or provided to others without their agreement. Banking and credit card systems often have the longest, most detailed, and onerous terms of conditions. Late payments on credit card bills can hike interest rates from 7 to 23 % if the bill was delayed or never received in the mail. There is at least a suspicion that sometimes bills and statements are complicated in unnecessary ways so that fees and other charges go undetected. Once one is signed up for automatic payments it is frequently hard to undo the arrangements, and annual credit card dues are charged without a billing process. Industry arrangements for online payment are convenient, but there is also a price to pay.

## Individual Transgressions and Cyber-World Concerns

This leads us to what individuals should be doing to look out for their own interests. The following are some good rules of thumb that suggest some things that are important to observe in today's complicated cyber-world. It is a complex and difficult world. Cybersecurity starts with your personal vigilance. You need to begin by getting the type of cybersecurity outlined in the previous chapter. But this is really only the start. Here are just a few words of cyber advice that you should always observe.

- Be very careful about giving up vital information such as credit card numbers, Social Security numbers, banking account numbers, and passwords related to financial records or vital data.
- Recognize that automatic payments systems such as "Apple Pay," "Visa Mobile Wallet," etc., are convenient, but these are still early days, and security flaws have been detected in these systems. If you use them monitor your statements carefully against fraud or cyber-theft.
- Consider carefully which accounts you sign up for automatic payment and recognize that there are "costs" associated with such arrangements, such as hidden fees, annual automatic charges, and the possibility of undetected accounting errors.

- Never, absolutely never, will any reputable bank or agency ask you for your password online. It will perhaps ask you to reset it if you request a new password, but it will never ask you to divulge your password.
- Likewise, no agency, such as the FBI, a police department, the Social Security agency, your bank, a credit card company, or a taxation unit will ever contact you online about a crime, a payment delinquency, or other serious matter electronically—even if the e-mail address seems legitimate. The fake address is created through so-called pharming or phishing scams. If you are in doubt contact the organization or business directly by phone or in person. Never respond online to what seems to be a fake e-mail address.
- If you lose a credit card or it is stolen, you need to remember exactly which accounts might be linked to automatic payment on that card and notify the company of the new card number. In short, if you do use automatic payments, it is best linked to a bank account and not a credit card.
- Credit cards that are shared with spouses, sons and daughters, or other relatives are another risky way to keep account of expenditures and errant charges. Someone should be able to know directly rather than second hand if a totally wrong charge has been entered. It is amazing how "free" trial offers translate into an ongoing charge that is very, very difficult to correct.
- If you see an ad for a free service or free trial offer, read the fine print carefully and then read it again. Ninety percent of the time there is a catch, and once you have given your credit card number to an unscrupulous firm, it may be hard to recover.
- Anytime you sign up for something online, or even go to a website, you have probably given away your online identity. In order to protect your identity consider setting your computer so that your "cookie" cannot be captured.
- A large number of applications on your smart phone are subject to black hat hackers. Recognize when you use your cell phone, I-Pad, or other mobile electronic device, which you may be subject to eavesdropping by means of an electronic scanner.
- Paying bills by check may seem old fashioned, but you do not have the same risk of electronic interception.
- In today's world a techno-terrorist can attack in many ways that can endanger you and your family. Every household should have a "go bag" for emergencies; these supplies would include at a minimum a knapsack, first aid kit, drinking water, food, a compass, a portable wireless radio, and a knife.

# Conclusions

The rate of change in our world is incredibly swift. It would be easier for Adam and Eve to come and live in the world of George Washington and Thomas Jefferson than someone from the U. S. colonial period to come and live in today's world of cell phones, computers, air and space travel, satellites, 3-d printers, automatic weapons, artillery, tanks and chemical, biological and nuclear weapons. And that is not to mention "cultural and social advances" reflected in Spandex, breast implants, R-rated movies, and nude beaches.

The social, cultural, and economic environment of the cyber world and the twenty-first century is one in which the rate of accelerating change is itself speeding up—a condition that physics would describe as "jerk," and mathematicians would call a third order exponential. Even a decade ago terms such as bitcoin, smart phone, 3-D printing, 4G LTE, Twitter, social media, texting, Internet of Things, Snapchat, blogging and "showrooming" would likely get a blank stare. Not only is the world we live in changing, but it is increasingly hard to stay abreast of the technology, the terminology, and the social conventions. The rush to stay hip and current in our techno-world has more than its share of dangers.

As there are more and more people connected via electronics and dependent on service jobs that depend on networking and the Internet, our vulnerabilities to cyber-crime and cyber-terrorism grows apace. If there is a unifying theme to this book it is that it is okay to not be on top of the latest trend and fluent in the latest techno-babble jargon. The Internet is useful as a communications media and can offer convenience and an economical way to obtain information, products, and services. Nevertheless it can also be a dangerous venue in many ways. Your children can be exposed to pornography or dangerous social, cultural, or political views—if not at your house then at the house of friends and neighbors. There really is such a thing as too much information. Just because information is online it is not necessarily true. Much of the information on the Internet is out of date, inaccurate, or even just plain wrong. It is useful to get information from reliable and trusted sites or rely on direct information from your doctor, druggist, banker, stock broker, or teacher/professor (Fig. 3.3).

One of the current problems in education, training, and critical thinking is what is quite correctly called "information overload." Sociologists Robert Merton and Paul Lazarsfeld experimented with laboratory rats and what happened to them when exposed to too much information. In the experiments the rats were exposed to music, white noise, strobe lights, changing

**Fig. 3.3** One of the big challenges of the cyber age is coping with too much input or 'information overload' (Graphics from author Joseph N. Pelton's book *Future View*.)

shades of colors, radio programming, and other stimulants. The results were not good. Over time the rats became apathetic, lost their appetite, stopped exercising, ceased eating, lost their sex drive, and eventually died. Too much information can be as large a problem as too little. Over dependence on the Internet can actually turn into an addiction. The State of California actually has an addiction center to treat people that have become addicted to the Internet. These are people that have lost their jobs, their spouses and children, and everything else. They literally were not able to go to the bathroom, eat, or sleep properly due to their addiction [14].

The issue of rapidly accelerating information technologies must ultimately lead to philosophical questions. The current estimates that the information available to humans is now doubling about once a year, and stored electronic information is not now measured in gigabytes or terabytes but in exabytes (this is to say, not billions or trillions but rather quadrillions of bytes of information). This has led to speculation as to where does this incredible rate of information accumulation and processing ultimately lead the human race? Do we become smarter? More servile? More dependent and thus ultimately servants to our "smart" electronic machines?

Professor Marshall McLuhan speculated on this in his writings on the creation of what he called the Electronic Village. Arthur C. Clarke in his seminal science fiction novel talked about "the breakthrough" in which a latest generation of humans morphed together to create one vast interactive global brain. In earlier books such as *Future Talk* and *e-Sphere: The Rise of the World Wide Mind* the question was posed as to what would be the meaning if a new sort of human called "homo electronicus" might mean for humanity in its ultimate evolution. Might 1 day humans be able to dispense with cell phones and communicate via alpha waves from our brains? Someday cyber technology may liberate us and make us smarter and better, or perhaps it will make us a slave to our technology. Such speculation is for another day, but for today, humans should try to protect their privacy, protect their personal identity, protect their passwords, and protect their intellectual and financial property by following the advice offered above (Fig. 3.4).

Such questions about our digital future and where electronic technology is leading are heatedly debated within political, business, economic, social,

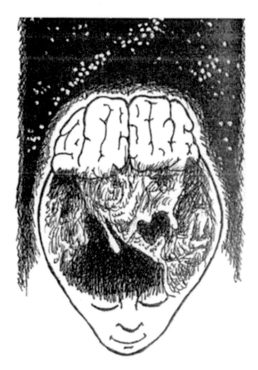

**Fig. 3.4** The ultimate future? Are we headed in a data intensive world: A global brain? (Graphics from author Joseph N. Pelton's book *Future View*.)

cultural, and religious circles around the world today. There are no definite answers, but the principles of privacy, liberty, freedom, and even the less well defined "the pursuit of happiness" seem valid concepts for whatever future world we create for human existence. One can hope for governments, industry, and individuals to agree on those principles.

# References

1. Tom Risen, "FISA Judge Denies Surveillance Court Offers 'Rubber Stamp': Surveillance Court and NSA face scrutiny on email address collection" *U.S. News & World Report,* October 16, 2013 http://www.usnews.com/news/articles/2013/10/16/fisa-judge-denies-surveillance-court-offers-rubber-stamp.
2. Lichtblau, Eric (July 6, 2013), "In Secret, Court Vastly Broadens Powers of N.S.A." *The New York Times.* Retrieved July 9, 2013.
3. Alison Smale, "Anger Growing Among Allies on U.S. Spying" *The New York Times,* Oct. 23, 2013. http://www.nytimes.com/2013/10/24/world/europe/united-states-disputes-reports-of-wiretapping-in-Europe.html?_r=0.
4. David Meyer, "UK terror law amendments would bring back 'Snooper's Charter.'" Jan. 22, 2015. https://gigaom.com/2015/01/22/uk-terror-law-amendments-would-bring-back-snoopers-charter/.
5. "Europe's Cyber Security Policy Settings Under Attack" *Space Daily,* May 4, 2014. http://www.spacedaily.com/reports/Europes_cybersecurity_policy_settings_under_attack_999.html.
6. Keven Bankston, "West Coast vs. East Coast: The crypto cold war between the feds and the tech industry just got hot." *Slate,* February 25, 2015. http://www.slate.com/articles/technology/future_tense/2015/02/yahoo_s_alex_stamos_and_nsa_s_mike_rogers_fight_about_encryption.html.
7. *Ibid.*
8. James Bamford, "NSA Is Building the Country's Biggest Spy Center (Watch What You Say)" *Wired,* March 15, 2012. http://www.wired.com/2012/03/ff_nsadatacenter/.
9. Securing Cyberspace – President Obama Announces New Cybersecurity Legislative Proposal and Other Cybersecurity Efforts, White House Press release, Jan. 13, 2015. https://www.whitehouse.gov/the-press-office/2015/01/13/securing-cyberspace-president-obama-announces-new-cybersecurity-legislation.
10. Where to get PGP(Pretty Good Privacy). www.cryptography.org/getpgp.htm.
11. Ellen Nakashima "Congress Gets Moving on Long-Sought Cyber Bill" *Washington Post,* March 25, 2015 P. A-6.
12. "Pressure to Rein in NSA Data Collection" *Washington Post,* March 25, 2015, P. A-6.
13. U.S.A. Freedom Act Passed, USA Today, June 2, 2015, http://www.usatoday.com/story/news/politics/2015/06/02/patriot-act-usa-freedom-act-senate-vote/28345747/.
14. Joseph N. Pelton, *e-Sphere: The Rise of the World Wide Mind.* (2000) Quorum Books, Westport, Connecticut.

# 4

# Ten Key Things to Protect

## Introduction

When it comes to protection of your physical assets, the approach is relatively straightforward. You can install a home security service in your home with burglar alarms, fire detectors, plus carbon monoxide and radon detectors. You can have an security service install smart monitors on doors and windows and even have laser beam detectors that can be turned on at night to detect intruders. You can also install a safe, and if you have the resources you can install a secure "panic room" equipped for you to stay safely for days if need be. It was revealed that an Arab sheik with a multi-million-dollar condo in New York City had installed three "panic rooms" in his $50 million complex. Most of us know how to set up at least rudimentary security systems for our home. Most of us can't afford and don't need a panic room, but the steps outlined in this chapter can help minimize the damage that might be done by various types of cyber-attacks.

When it comes to automobiles, safety research continues to bring new technology to bear. A modern car will likely have an automatic braking system (ABS), automatic door locks, perhaps eight airbags for front and side collision protection. Then there are safety deposit boxes at banks for jewelry, key documents, and unlaundered drug money (just kidding!).

In the world of cyberspace and intangible assets such stocks and bonds, lines of credit, passwords to one's banking accounts, private e-mail accounts, cell phones and text messaging, Social Security and Medicare accounts, and literally over 20 different electronic accounts and protected websites, the best path to personal cybersecurity most certainly involves two concepts. First, you must know that it is difficult to protect your privacy and your

© Springer International Publishing Switzerland 2015
J.N. Pelton, I.B. Singh, *Digital Defense*, DOI 10.1007/978-3-319-19953-5_4

accounts without it being a hassle and burden. Secondly, it is very likely that in the fast-changing digital world, where information flows are increasingly broadband, global, intricately networked, and complex, security procedures and network protections will likely become more complex and subject to frequent updating and change.

In short, the matter of cybersecurity is likely to be hard, potentially complex, and fluid! It may also spill over into other key parts of your life, such as financial planning, your life's legacy, and even your last will and testament.

For those with small business and large financial assets you may well consider obtaining a personal cybersecurity consultant. Having such an easily available advisor may become yet another service you will need to obtain. Just as people have personal trainers, financial advisors and accountants, personal shoppers, dog walkers, and child care giver, you may very well wish to retain a cybersecurity advisor. If you do decide to do so you need to make sure that his or her credentials are faultless and fully vetted before relying on them. This means checking references from people you personally know and trust. Such advisors should most likely be obtained through your bank, stock broker, or other reliable source rather than just locating them online, where scammers can lurk.

## The Top Ten Things for You and Your Family to Protect Against and How

There are more things to protect against black hat hackers than you might think of right off the top of your head. Further, you might feel your personal key assets are adequately protected, but there are others to consider as well. You may have elderly parents or loved ones who are not the most computer literate and need assistance in making sure their cybersecurity is as strong as it should be. Also others may be quite computer literate but may still not be well versed in all financial matters related to retirement planning, stock market or real estate investing, or even budgeting to take account of life's reversals such as a corporate layoff when a "right sizing" from the top unexpectedly occurs at their place of employment. Consider the following advice in terms of not only yourself but also in terms of loved ones who might need better cybersecurity or enhanced assistance with financial planning.

## #1. Protect Stocks, Bonds, and Retirement Accounts

Most people today have stock and bond holdings whether they know it or not. This is because the great majority of people have Individual Retirement Accounts (IRAs) (either Keogh, Roth or both), 401(k)s, or other retirement plans through one or more companies or agencies. These stocks and bonds that are invested in the company itself through a 401(k) or through various investment funds. Money is currently being made available by national banks (such as "the Fed" in the United States) to commercial banks at, or close to, historic lows with near zero percent interest. Because of these low interest rates, investments for retirement accounts have migrated to stock equities, mutual funds invested in equities, exchange traded funds (ETFs) that focus on a particular market sector, or to various types of real estate investment funds. Assets that are taxable might well be invested in tax-free municipal bonds. If these "munies" are purchased within a local jurisdiction they can also be free of state taxes as well. Some corporate bonds can still pay reasonable yields, but most of those paying high yields are rated to be highly risky and are appropriately known as "junk" bonds.

Retirement accounts are not subject to taxation, until withdrawn. In the year when one turns 70.5 in the United States there are so-called required minimum distributions (RMDs). The amount of these funds that must be cashed out and reported as income for income tax purposes is based on a complicated formula that starts out with you having to liquidate the account within 27.4 years starting at age 70.5. If you live to be over 100 it shrinks down to 3 years. If you have a good investment counselor you can make back much of the funds represented by your minimum distribution by the time you have to cash out another chunk of your retirement funds the following year [1].

In addition to protecting your IRA account from hackers you also need to give it some attention, and there are ways that this account can be leveraged, used in real estate transactions and loans, etc. Thus, what we are saying is that it is important to manage your IRA actively as well as to protect it [2].

Protecting your IRA or 401(k) is key, though. This may be the most important asset you have. You really do not want to keep your stock or bond certificates at home, in a safe or a safety deposit box. It is best to keep these with a trusted stock broker that is competent and with a well-established firm. A competent broker will have you invested in a diversity of funds. In some cases you can select from a series of funds and invest in individual stocks yourselves. We recommend that you do not personally invest more than 10 % of your assets and leave it to professionals.

**Fig. 4.1** Buy equity stocks only with a competent and cybersecure broker

There is pending legislation that seeks to protect retirees from stock brokers that encourage investment in funds with high fees and low returns. Do your own research to find the most competent and reliable investment firms and which firms to avoid. In many cases you will have worked for more than one company and then you may actually have stock and bond certificates with two or more brokers. This diversity is an additional protection and thus is an asset in terms of risk reduction (Fig. 4.1).

The main thing is to let your stock broker, investment fund manager, or corporate retirement fund manager be the safe custodians of your certificates. This would include tax-free municipal bonds, where you invest your savings that are subject to be taxed. Also make sure that you make and store printouts of your investment holdings at least once or twice a year. Keep these records in a safe place. A reputable nationally known stock broker will keep its records backed up at several safe locations and also store these records in a "Cloud" memory service that is also protected in a variety of ways. Nevertheless a paper copy of your assets and net valuation is a good thing to keep in a protected file. Monitor the records of all your holdings to make sure there is no significant change in value or losses that are not easily explainable. Do not be hesitant to call your broker or investment fund if there are numbers that do not add up. Make sure that there is an electronic monitor site that can allow you to at least check on results ever 2 weeks or every month—at a very minimum.

To recap follow these rules: (a) Make sure you are working with a trusted stock broker or investment firm that provides clear and complete records that you can monitor via a well-protected electronic site; (b) Let these trusted professionals keep all stocks, bonds, and real-estate investment certificates and records. Even if you should acquire some stock with certificates provide these to the professionals, and they will add these to your records. They are safer with the firm and easier to trade as circumstances arise; (c) Keep a paper copy (hardcopy) record that is no more than 6 months out of date at the maximum; (d) Make sure that you have very good passwords and account IDs that are changed regularly and that these passwords are hidden and different from the password you use for accessing a public library or pizza delivery shop.

## #2. Protect Your Bank Account and Credit Card Account Records

Have a checking account, a savings account, and an overdraft protection account that automatically issues a loan if a check should somehow overdraw on your checking account. If you should actually have such an overdraft, the bank will notify you. This you should promptly pay off, but this short term loan will, at most, cost a dollar or two and will be much less than a $30–$50 returned check fee plus you will avoid potential damage to your credit rating. As discussed earlier, you should also limit the number of credit cards that you have.

Deals that lure you to open up a new credit card such as 50,000 airline miles, or a 5 % discount on purchases, or a toaster oven, are perhaps appealing, but if you have eight to ten credit cards in circulation, this hurts your credit rating and raises questions as to why there are so many different lines of credit. This can be a ticklish issue within a family between a husband and wife or between parents and children. You need to explain to all concerned that financial advisors STRONGLY recommend taking out no more than three or four credit cards. Don't get a new one just because a university or noble cause asks you to take out a credit card. Make a donation to them instead. This will be a tax write off, and you will have improved your credit rating and protected yourself against a possible future fraudulent charge against your account by a cybercriminal. It is not a good idea to have too many bank accounts or too many credit cards. If you have your credit cards with your local bank or stock broker who knows you this can work to your advantage over time. Also follow the guidelines. Call your broker or banker if you are going to be traveling, and change your pin number. Insist on getting a credit/debit card with a smart chip that is a further protection against black hat hackers getting your credit card number and PIN number and selling it to a professional cybercriminal who counterfeits your card and goes on a shopping spree. If your bank is not issuing credit or debit cards with an EMV chip you should consider moving to another bank. Here U. S. banks are years behind European banks [3].

In time one can hope that there will be upgraded security systems with banking cards and credit cards that will only be activated with facial recognition software or otherwise be better protected than current banking and credit card systems are.


## #3. Protect Your Social Security, Medicare/Medicaid Accounts and Medical Records

After your bank accounts and your stock and bond accounts (particularly retirement accounts) the other types of records that you should most want and need to protect are your Social Security account, your medical records and your Medicare/Medicaid accounts. These unfortunately are very desirable prey for cybercriminals. Fortunately these records are rather stringently protected. On the other hand, these are accounts that if compromised you might not know it until considerable harm has been done.

As noted in other parts of this book, it turns out that cybercriminals pay a premium for medical records. This is because medical billings tend to come in large chucks. Thus a bogus charge for something like $11,900 for a surgery, or $2100 for an automated cart, or $7000 for a prosthetic device would not prompt particular notice. Banks and credit card companies have computer programs to alert them to what would appear to be an irregular range of charges, but large medical charges are not unusual when they occur. The cybercriminals who specialize in trying to collect bogus claims against Social Security accounts, Medicare or Medical accounts are among the more sophisticated practitioners, because they are subject to being tracked down by the FBI due to a complex web of federal laws.

Your best protection in this case is to choose a well-crafted password for your Social Security and Medicare accounts and ask your various doctor's offices about their security procedures. If they use online reporting of medical tests they must be very careful in their use and opt for telephone dial-in reporting instead if that is an option. Medical offices will provide you with privacy statements, but unfortunately these are long and complicated. Insist that they walk you through any forms that they ask you to sign before putting your signature on it. In the age of automation, protection of one's privacy is one of the hardest individual rights to maintain. Yet privacy rights are at the core of democracy and basic human freedom [4].

## #4. Protect Desktop Computers, Wi-Fi, and Computer Routing Systems (LANS and WANS)

The chapter that follows spells out where to go to get assistance and provides a practical guide to things you should be doing to protect your cyber-security for your desktop computer as you connect to your Internet service provider, use Wi-Fi hot spot connections, or utilize computer routers in your home or office's local area network. Even corporate-wide area networks, enterprise networks, or virtual private networks (VPN) that you are connected to on a personal or professional basis can be intercepted. Unless you obtain protection in terms of an antivirus, a firewall, and insurance against identity theft you are likely to hacked with a variety of bad results. All wireless connection nodes should be protected with a secure password. Wireless routers are extremely convenient and support mobile access, but they are also vulnerable. Any other scanner in range can access the signal, and without password protection and encryption your messages become susceptible to interception (Fig. 4.2).

Your vulnerability could be as trivial as unwanted spam, to much more serious things due various forms of malware. There could be financial losses, loss of personal or professional information that you wish to protect. You might be subject phishing or pharming schemes that result in you giving up vitally important unprotected information. You might be infected with a Trojan, data bomb, or virus that puts you at risk of having to pay a ransom

**Fig. 4.2** Wireless routers to support a home or office local area network are convenient but are subject to cyber-attacks

to allow you to recover the use of your computer and all the information that is stored on it. Such ransomware is just the latest form of computer harassment.

Just to check on my own computer antivirus I looked at the report for the last 30 days. My File Systems Analysis found two threats caught from 98,000 files scanned. The Web Shield Analysis found five threats in 267,000 web objects scanned and that 174,000 new Virus Database Updates had been added—all in just 1 month.

The most serious computer-based challenge today is identity theft, where not only could you have money or assets stolen from you, it is possible that you could be charged with felony offenses in the form of computer crimes that it appears were committed by you. Wireless connections are the most vulnerable. Public Wi-Fi networks should be used only to get information such as bus schedules or office addresses or to summon an Uber cab, but never to conduct financial business or convey confidential information. Even wireless router systems in offices (i.e., wireless LANs, MANs and WANs) are vulnerable, and unauthorized users can get access. Read the following chapter carefully and sign up for antivirus, firewall, and identify theft protection at a minimum. If you choose the best and still economical anti-malware systems you can get a reasonable level of protection for around $60–$70 a year. This level of protection is a wise investment.

Wide area networks (WANS) used by companies to interconnect teleworkers can be extremely vulnerable. It takes only one worker accessing a protected network on a non-code protected wireless LAN to access critical corporate information by the simple use of a scanner in a car near the home of the thoughtless telecommuter. Unless you are working on a computer completely offline, you should consider almost anything that you type or view online to be publicly visible. Unless information is digitally encrypted there is little ensured privacy in today's world.

## #5. Protect Personal Cell Phones/Smart Phones

Just as wireless LANs and satellite-based virtual private networks are a security concern, where IP security (IP sec) can be violated, one has to be very careful with smart phones. Most people think that if they have good passwords that they can be safe using their smart phones for banking and other private activities, but the truth is that cybercriminals can use social media,

Wi-Fi connections, wireless LANs, and scanners can extract the most sensitive of information. One can now use face recognition to access I-Phones and Android smart phones, which is better protection than a password.

One should use smart phones to communicate or acquire non-sensitive information that you would not mind sharing with the rest of the world. As we enter the world of Apple Pay, Xoom Money Transfer, Bitcoin, and other electronic money systems, there is the danger of cybercriminals finding ways to using phishing and pharming schemes to commit cyber-fraud. The various systems such as Apple Pay, LoopPay, Google Wallet, and Plastic and Coin use different technologies. Apple Pay uses near field communications, LoopPay uses magnetic secure transmission, and Google Pay and Plastic both use an Android iOS system that also requires near field communications, while Coin use a magnetic strip reader like Loop and is not compatible with the EMV chip readers that most credit card companies are now issuing. The point is that the magnetic strips are clearly vulnerable to credit card fabricators with stolen credit card information, and near field communications is far from foolproof [5].

Of course there is the reverse proposition of the use of cell phones—not to steal your money or private information, but rather to track you and your activities. In Pakistan, the government is requiring all cell phone users to provide biometric information (i.e., fingerprints) linked to their cell phone/subscriber identity module (SIM card) in order to combat terrorism. In the United States drug smugglers are using so-called untraceable "burner phones" or even stealing a car to use the cell phone inside to confirm a drug deal [6].

## #6. Upgrade Cyber Security Systems for Vital Infrastructure

It is not possible for an individual to secure vital infrastructure against black hat hackers and techno-terrorists, but it is possible to be vigilant and encourage responsible governmental action to provide increased protection. At the local level there should be a commission that looks into the security of local governmental databases, the security of so-called "SCADA—Supervisory Control and Data Acquisition" systems that control traffic lights, water and sewerage, power substations, and other vital infrastructure. Such a review in our local jurisdiction led to a new security upgrade and independent audit by an outside expert on an annual basis.

At the state, regional, and federal government level, there are new standards for the security of information networks that control power substations, nuclear power plants, railway and airplane traffic control systems needs strong public support. Most recently in the United States a report from the Government Accountability Office (GAO) concluded in March 2015 that: "The Federal Aviation Administration has fallen short in its efforts to protect the national air traffic control system from terrorists or others who might try to hack into the computers used to direct planes in flight." The GAO recommended 14 specific steps that should be taken to upgrade airline traffic control networks against cyber-attacks. When such reports are released a vigilant public needs to send messages to their legislators and other public officials voicing concern and calling for greater vigilance against cyber-attacks [7].

This is not only a matter of local, state, and federal government vigilance and protective actions using encryption and secure codes against computer network attacks. It is also a matter of taking protective actions by corporations that provide vital infrastructure (i.e., power, water, transportation and communications) and systematically protecting against such cyber-attacks as well. The trouble is that there are always more and more sophisticated ways that cybercriminals can attack vital infrastructure, but digital encryption and division of networks into segregated parts can limit the damage, and hopefully foil cyber-terrorist attacks [8].

## #7. Protect Big Data and Social Media: And At Least Some of Your Privacy

Part of the problem is that people and businesses are promoting the creation of a more cyber-vulnerable world that is ever harder to protect. The world of social media, where younger people are sharing more and more information about themselves, increases their vulnerability. Corporations constantly processing what they call "big data" about consumers that can be found through their use of the web and their cell phones. As author Marc Goodman in his book *Future Crimes* has said; "People are not Customers, They are Product." In a world in which computer processors can store an amazing amount of information about you and analyze your likes and dislikes, there is little privacy, little that can be hidden, and much that is exposed and vulnerable [9].

In short if you do want to protect your privacy, your bank accounts, your stock trades and bond purchases, plus real estate purchases and your credit card transactions, it is certainly wise to be careful and "electronically prudent," but short of going to a desert island, or having someone operate as your agent and never going online, there is little that you can do to protect your privacy in any absolute way. The key is to at least be better protected than others, embrace digital encryption or other techniques such as facial recognition network access, and encourage governmental action to increase cybersecurity processes and enforce stronger cyber protection laws [10]. It is probably a good idea to not tell everything about yourself to the world over social media. A Twitter account of how you get serially drunk, or a Facebook posting of you topless at the beach or at a beer bash is probably a bad idea. This might not lead to a cyber-attack on your bank account or identity theft, but it may cost you that job that were hoping to land.

## #8. Don't Put All of Your Eggs in One Basket

Most people have their financial assets rather automatically distributed in terms of real estate, tangible property and vehicles, a bank account, a savings account, a retirement account or a Social Security and Medicare account. They may have an investment firm as well. If one works for several firms during their career they have retirement accounts with several firms.

Personally one of the authors has investment accounts with Schwab, Merrill-Lynch/Bank of America, and TIAA-CREF. There is often pressure to "roll over" one's savings or retirement accounts to be with one company. There are several reasons why one should think twice about such consolidation. First off, in the case of savings accounts—as opposed to retirement accounts—it is important to know and have a record of when stock was purchased and at what price because this is significant for capital gains tax. You cannot take a loss or have a clear record for capital gains if you shift your investment accounts around, because your records are deleted after you drop your account. The second reason is that different financial advisors have different investment strategies, and you can diversify your risk and compare performance. The third and most important reason is that if one firm is attacked by a cyber-criminal, or perhaps fails, you are diversified, and insurance protection by the FDIC only goes up to a certain limit with each account.

In the case of credit cards or debit cards, the logic runs in reverse. Having more than three credit or debit cards is not advised, because the more cards you may have the greater the chances are of that company having their records hacked and thus ending up with you being exposed to fraudulent charges. When it comes time to pay income taxes, your records will be somewhat more complex, but the diversification of the risk is worth putting up with the complexity or paying a CPA or tax accountant somewhat more for having one or two more accounts.

## #9. Protect Yourself by Doing Audits

Business and organizations have audits each year to make sure that their financial accounts are in order and things are as they should be. Today it is increasingly important for organizations to have their computer accounts and operations (i.e., secure data storage systems, SCADA networks, secure communications, and IT networks) audited to make sure that there are no bugs in their network (i.e., malware, viruses, data bombs, unauthorized "certs" in wireless LANs, and Trojans) and that they have not been subject to so-called phishing and pharming scams. Data system managers like to think of themselves as highly capable and immune from cyber-attacks, but an independent outside review of all aspects of communications, IT networking, and emergency power backup systems should be carried out by an outside expert who is aware of the latest types of cyber-attacks and able to examine all systems and procedures with a fresh eye to any irregularities that might exist.

In companies, governmental agencies, and even families, it is sometimes hard for individuals to admit that their practices are in need of improvement, especially when the critique comes from close at hand. An outside consultant who is expert in cybersecurity and aware of the latest cyber-attack technology and system can very useful for reviewing the situation. This may be as simple as getting a colleague who is an IT expert to come in and review a wireless LAN in a home, or home office, and to see that vital files are backed up on a flash drive, or tax files and financial records are stored in a safety deposit box or fireproof container. On the other end of the spectrum, it may involve hiring a firm such as Accenture, Deloitte, Los Alamos Technical Associates to do a full-scale audit of all cybersecurity and IT operations to see what practices should be changed or upgraded.

### #10. Protect Your Personal Things, Your Family, and Your Business Interests, Including Your Last Will and Testament, Plus Donations and Legacy

The previous nine points note some of the things that you need to protect yourself against cybercriminals and potentially even cyber-terrorist attacks. A terrorist is not likely to attack you personally, but they may launch an attack against your community's vital infrastructure that could create a life or death threat to you and/or your family. The steps outlined in this and other chapters of this book can make you more secure and less at risk to a cyber-attack. There is a tendency today to put almost all vital information, including family photos, key documents, and important correspondence online. It is important to recognize that computer data is not necessarily permanent. Even things stored on the web are not necessarily permanent. It is important to have paper records of vital documents such as birth and marriage certificates, passports, real estate deeds, historical photographs and family trees, and other mementos in a safe place. You should, at a minimum, protect things like important family photos and historical documents. These should be scanned and then put on flash drives or on an external hard drive and stored in a safe place not in one's home but in a safety deposit box. Absolutely everyone should have a living will and a legally binding last will and testament. If one dies without these, it is putting a big burden on loved ones, who will have to deal with a probate court for perhaps over a year to make decisions that you should have made yourself.

Especially documents such as wills should be kept on file at the local Clerk of Courts, with law offices, if drafted by an attorney, or on electronic file if prepared by an online service such as Legal Zoom. This secure handling of vital documents also applies to a living will, to final instructions that involve the provision of donations to key charities, and perhaps things that you wish done to preserve your legacy or the legacy of your family and children. It is not uncommon for people to think that they will last for a long time and put off making these sorts of arrangements, but it is prudent to make sure that everything is spelled out in formal written documents that are witnessed and on file with appropriate officials. There are many things that need to be in electronic form and are efficiently kept there. Part of this preservation process is to not lose things in a fire, hurricane, or through death. The other reason is that this is also a form of protection against various types of cyber-criminal attack. It is unfortunately true that some cybercriminals are attuned to death notices and seek to take advantage of accounts that might be particularly vulnerable to attack. Just a few weeks ago a spam message, perhaps containing a virus, came in via e-mail from a colleague that had died several weeks earlier.

## Conclusions

It turns out that the practical steps that one might take to defend against cybercriminal and cyber-terrorist attacks are those that make sense for other reasons as well. Today cybercriminals can attack your bank accounts, credit cards, stock and bonds holdings, real estate investments, or even use information that is stored on your social media accounts. What you post on your Facebook or Linked In page could indeed be used to steal your identity. Some of the attacks do not necessarily involve digital technology per se, but they are just made easier due to the Internet and computer messaging. Some years ago, while serving on a grand jury in Northern Virginia, there was a case of a man who married a Vietnamese woman who owned a home. This person then managed to get a loan on the home and then made a few payments without her knowing and using a box office address for the transaction. He then disappeared with the proceeds from the loan. The point is that this is a type of fraud that could have been carried out prior to the Internet. But today, with phishing and pharming activities, it is much easier to defraud people using the Internet and texting and networking schemes and commit the fraud with anonymity and speed.

Few people today fall for the Nigerian Prince fraud, but there are dozens of scams to take the place of the earliest attempts to commit computer fraud. The chapter that follows provides a great deal of practical advice as to how to protect yourself and your computer. There are, however, only a few logical rules to follow that are always very helpful to remember.

- First, if it sounds too good to be true, it probably is.
- Second, no bank, financial institution, university, utility company, etc., will ask you to divulge your password via a computer inquiry or over the phone or via a text message. Don't do it!!!
- If you get a message that appears to be from a bank, a government agency, or a utility company and seems to come from a legitimate e-mail address but makes some unreasonable claim of payments or tax garnishment, please recognize that this is likely a pharming scam that tries to make you think it is legitimate. Check directly with the organization in question, but not via the e-mail-delivered information.

• Get the type of digital protection we outlined in earlier chapters. This includes an anti-virus program, a firewall, anti-identity theft protection, and if appropriate sign up with an anti-cyberbullying service. Also have an access-protected and GPS tracking capability on your smart phone, and make sure that applications that you download to your smart phone are not secret Trojan horses.

# References

1. "What Is My IRA Required Minimum Distribution?" *Kiplinger,* January 2015. http://www.kiplinger.com/tool/retirement/T032-S000-minimum-ira-distribution-calculator-what-is-my-min/index.php.
2. Publication 590, Cat. No. 15160X, Individual Retirement Arrangements (IRAs), Internal Revenue Service, U. S. Department of Treasury. http://www.irs.gov/pub/irs-pdf/p590.pdf.
3. Cathleen McCarthy, "U. S. rolling out chip card technology, ever so slowly." http://www.creditcards.com/credit-card-news/us-slowly-rolls-out-emv_chip-technology-1276.php#ixzz3TNIGIoKR.
4. Jeff Stone, "Chinese Hack of 4.5 Million Hospital Records Is 'Identity Theft On A Platter' For Cyber Criminals," August 19, 2014. http://www.ibtimes.com/chinese-hack-45-million-hospital-records-identity-theft-platter-cyber-criminals-1663066.
5. Jason Steele, "Four Alternatives to Apple Pay," February 26, 2015. http://finance.yahoo.com/news/4-alternatives-apple-pay-110054904.html;_ylt=AwrBEiRh4_RUME8AJRzQtDMD.
6. Tim Craig and Shaiq Hussain, "Pakistani's Choice: Verify Identity or Forgo Cellphone," *Washington Post,* Feb. 24, 2015, p. A1 and A9.
7. Ashley Halsey III, "FAA Computer Systems Vulnerable, Report Says," *Washington Post,* March 3, 2015, P. A2.
8. Marc Goodman, *Future Crimes* (2015) Random House, New York.
9. *Ibid.*
10. Omer Tene & Jules Polonetsky, "Privacy in the Age of Big Data, A Time for Big Decisions," *Stanford Law Review,* February 2, 2012. http://www.stanfordlawreview.org/online/privacy-paradox/big-data.

# 5

# Cybersecurity for Smart Phones, Mobile Apps, and "The Cloud"

## Introduction

The world of cyberspace was for quite a few years largely confined to that of the desktop personal computer. These connections to the Internet could be reasonably well protected by installing anti-virus and firewall protection against those early black hat hackers or crackers that might seek to install malware into your computer.

We are now in the new age of prevalent wireless connections to the Internet. To live and operate in the world without a smart phone and I-Pad and other wireless paraphernalia is increasingly difficult. Even public institutions such as airports, train stations or subways stops are largely devoid of pay phones (at least any that work) because everyone is just "assumed" to have a cell phone. Bus stops or train stations no longer feel they need to post schedules since everyone can just go online to check times and prices. Even checking in a airport today is largely a matter of dropping off luggage, since a growing percentage of people select their seats and print out their boarding passes online. The convenience and ubiquity of small cell phones is clearly changing our world. God help us if we lose a smart phone, a charger, or a battery runs down.

But there is another challenge beyond having the latest model of smart phone. They are likely carrying around a vulnerability about which many people are largely unaware. It is today hard to get through a day without the apparent need to download a new mobile application that has suddenly become essential to daily life. Newspapers, television news networks, television show and movie distributors, music download sites, airlines, even restaurants and fast food chains are vying for you to download their app. Luring people to download a commercial app is today almost like getting a

© Springer International Publishing Switzerland 2015
J.N. Pelton, I.B. Singh, *Digital Defense*, DOI 10.1007/978-3-319-19953-5_5

user hooked on a drug. If you have Hulu, Netflix, or other download apps on your smart phone, this is potential money in the bank for a sophisticated hacker.

The long and short of it is that when people download an app to their cell phone they have also likely expanded the problem of their Internet security and increased their vulnerability of a cyber-attack Today there are typically more users linking to the Internet through wireless systems than wired systems, such as coaxial cable and fiber.

Over the last decade there has been a tremendous growth in the use of "smart phones" (i.e., broadband 4G LTE cellular), public Wi-Fi systems, wireless LANS in homes and offices (i.e., wireless routers) plus expanded use of satellite services to reach locations not served by broadband fiber. Particularly as users install more and more applications on their smart phones or on computers supported by wireless routers there is a need to install new software to protect against various types of cyber-attacks as vulnerabilities have grown. There are now literally billions of users of I-Phones and Android phones that have added a dizzying array of new applications without suspecting the hidden cyber-security problems hidden in the apps they are making available.

Also there is the problem of accessing the Internet using public Wi-Fi systems that are not password protected, or even Wi-Fi networks that are password protected, but use a very low level of protection that is easily broken into by black hat hackers.

And these are just of the new and growing risks associated with mobile access to the Internet. Another cyber threat has come as many companies, banks, and public institutions have begun using computer processing power stored in "The Cloud." These commercial and governmental services on which consumers depend for a widening range of data storage and networking services can also be vulnerable.

Nor are these two areas of networking concerns the only problems. In the following chapter we will address the Internet security problem that is today posed by vulnerable satellite networking connections as well as the increasingly omnipresent, yet invisible, Supervisory Control and Data Acquisition (SCADA) networks.

Finally, the latest area of concern is the growth of the Internet of Things, which increases wireless vulnerabilities even more widely. Wireless connections to smart refrigerators, washing machines, baby monitors, and automobile fuel injectors will add problems as well in the next few years. An edition of the television show "Sixty Minutes" in March 2015 featured an expert presentation on how one could take over the control of a car using microprocessors that can be accessed via the Internet. In the future literally billions of "smart" electronic components will be accessible via wireless capabilities that are being built into vehicles and appliances to diagnose problems. Yet creating such a capability now represents a cybersecurity problem in itself. This newly emerging security issue of the Internet of Things will be addressed in the chapter on the future. Already today we are hearing about washing machines and refrigerators sending out spam and baby monitors being hacked. Tomorrow, every home and every vehicle could potentially be hacked for mischief, crime or even a terrorist attack.

## The Dangers of Wi-Fi Access to the Internet

Every time a new capability or convenience is added to our complex technological world the odds are that a new problem or risk will be added as well. Wi-Fi systems are very convenient to access the Internet when one is traveling or on the go, but there are definite risks that one should be aware of even if you are accessing a password-protected site.

Avast, which is a cybersecurity firm that was discussed earlier, recently conducted an experiment where they observed actual behavior in nine cities around the world (three in Asia, three in the United States, and three in Europe). This extensive survey revealed that a significant portion of mobile users browse primarily on unsecured HTTP sites. This worldwide usage survey found that based on their extensive sampling that 97 % of users in Asia connect to open, unprotected Wi-Fi networks. The lowest levels of access to unprotected were Barcelona and San Francisco, but at 80 % this was still very high.

The even worse news, for those who thought they were being cautious, is that some seven out of ten password-protected routers use weak encryption methods. This weak encryption makes it simple for most password-protected sites to be hacked. Nearly one half of the web traffic in Asia takes place on unprotected HTTP sites, compared with one third of U. S. traffic and roughly one quarter of European traffic. This can most likely be attributed to the fact that there are more websites in Europe and the United States that use the HTTPS protocol than in Asia.

So, how concerned should you be? How much of your browsing activity can actually be monitored on Wi-Fi wireless networks?

Because HTTP traffic (as opposed to HTTPS traffic) is unprotected, the AVAST team that conducted this global study was able to view all of the users' browsing activity, including domain and page history, searches, personal login information, videos, e-mails, and comments. Before the start of any communication, there is always an exchange with the domain name server (DNS). This communication is not encrypted in most cases, so on open Wi-Fi it is possible for anybody to see which domains a user visits. This means, for example, that somebody who browses products on eBay or Amazon and is not logged in can be followed around. Also, it is visible if people read articles on nytimes.com or CNN.com. Users who perform searches on Bing.com, or who visit certain adult video streaming sites, can also be monitored.

The majority of Wi-Fi hotspots such as those offered by hotels, coffee shops, etc., were password protected. Often these sites use so-called WEP encryption, which can easily be hacked. Using WEP encryption, according to the AVAST team's analysis, can be nearly as risky as forgoing password-protection altogether [1].

If there is a bottom line here it is to use public Wi-Fi systems cautiously. Never log onto bank accounts or stock broker accounts or any other sensitive site via these wireless networks where your password could be revealed or sensitive information obtained. If you are working from home, then if you are using a wireless LAN, it should be password protected with a high level of encryption—i.e., something more secure that WEP.

## Mobile Applications on Smart Phones

Some might argue that because there are more applications for the I-Phone, and because I-Phone users might be higher end consumers than Android phone users, black hat hackers will target I-Phones the most. Others, such as computer expert Eugene Kasperky of Kaspersky Labs, have argued that because Apple controls do not allow third-party applications, it is easier to protect I-Phones from cyber-crime than is the case with Android phones. Kaspersky has indeed argued that for this reason Android-based phones are significantly more targeted in cyber-attacks, because of vulnerabilities that come built into third-party applications. Kaspersky, however, also argues that there are very good systems, such as from Avast and 360 Mobile Security, that can scan Android mobile devices and find 99.6 or better of downloaded apps that present a security risk [2].

The truth is that there are potential vulnerabilities for all types of smart phones and that any time you choose to access the Internet via your mobile phone that certain precautions are in order. This is true regardless of which type phone or application you may use. It is also true that if you have a wireless LAN in your home or office, there must be a security concern as well. Anyone with a scanner can access a signal from a smart phone or a wireless LAN, and unless it is protected it is subject to interception.

## Mobile Phone Security and Access

One of the assumptions that people often make is that if they are the only one that can access their phone their cybersecurity can be assured. Certainly there are various ways to insure that your phone cannot be accessed by others using a simple numerical, letter, or symbol-based code. There are phones, for instance, that require you to draw a pattern, see the iris of your eye, or your finger print (Fig. 5.1).

There are also short messaging systems (SMS) or texting applications that can be used to track your phone if it is stolen or even to set off a loud siren noise that might encourage a thief to ditch your phone quickly rather than be apprehended. The problem is that this means that the thief might tend to throw it in a sewer, a lake, a river, or the ocean [3].

Such protection is quite useful, but the biggest security issue involves the "apps" you install on your phone.

**Fig. 5.1** Smart phone that requires you to draw a pattern to gain access to it

What is clear is that the most recent trend in cyber-attacks by black hat hackers has been to shift their focus from computers to cellular phones. According to Kaspersky, a virus similar in destructiveness to the Chernobyl virus of the 1990s could soon target cellular phones.

## Making Your Smart Phone More Secure

There was a sixfold jump in malware threats that infected smart phones during calendar year 2013 and another huge increase in 2014. This exponential increase in attacks against mobile devices was reported by Juniper Networks, the software company. As noted later in this chapter Android phones and installed apps can be particularly targeted because kits used to create these apps are subject to attack.

There is no way to totally protect your smart phone and the information stored on it, but here are some cyber protection strategies [4] (Table 5.1).

Table 5.1  Ten recommended ways to protect your smart phone and electronic devices

1. Enable encryption on your smart phone if it is provided by your vendor. This can provide you with some protection of your privacy.

2. If your Internet service provider or smart phone provider recommends a particular security software then proceed immediately to have it installed. There are several free products that can very usefully installed on your phone in order to help protect your phone. These can include AVAST (used by the author), Lookout Mobile Security, AVG, and others. These can also help with phone location if stolen. (See No. 7 below)

3. Be careful about installing any new apps on your smart phone, I-Pad, laptop, or desktop computer. These are the prime source by which cyber criminals can access your smart phone. Thus only buy apps from websites such as Google, Apple, etc. Also never save your password on any apps. This is a critical cybersecurity rule of thumb.

4. Be cautious about using publicly available Wi-Fi systems, and avoid banking or financial or stock market transactions on such sites. It is always better to use your home computer with a secure hardwire connection to your Internet service provider.

5. Don't leave any access connections to the Internet open, and turn off your electronic devices when not in use. If you have wireless chip in your cell phone or any electronic device others can access it with the proper air interface standard.

6. Clear your browser history on all electronic devices. This can be used by cyber criminals in a variety of ways, and this is yet another way to preserve your privacy.

7. Activate an application such as "Where's My Droid" or the I-Phone software to help you to always locate your phone and to help police find your phone if stolen. (See Number 2 above. Security software can also provide other protections as well.)

8. Back up your most important data on your desktop computer, smart phones and I-Pads either on a hard drive, high capacity flash drive or via a service that does it automatically on short intervals.

9. If your phone is stolen and not recovered, you can contact your carrier to remove sensitive data that was stored on it.

10. Lock your phone with at least a PIN number, but it is best to have an even more secure system such as facial recognition or where you are required to draw a design. (See earlier discussion in this chapter.)

# Cybersecurity and Banking Apps

The first place to start with mobile phone security is to recognize that there are security risks with "tap and go" or a "wave system" payments that do not require a credit verification method (CVM). There are several types of risks here. One is that your phone that contains payment information might be stolen and used by others. In some cases it could actually only be a matter of just getting your mobile phone number. There are systems such as that used by Facebook that might allow others who have access to your phone number to make purchases simply by placing an order using your phone number. The other risk is that the organization that has taken your credit card data might not have encrypted it, so that the information could be stolen. In the case of Starbucks, for instance, that was initially a potential problem [5].

The issue of contactless payment systems is of particular importance to address. The Apple Pay system is currently the most widely used in the United States, with many millions signed up for the service. Along with other providers Apple Pay relies on so-called near field communications (NFC) technology in order to allow instant purchases. Apple has claimed that their version of NFC payment is more secure than competitors because there is "tokenization of its data" in order to encrypt it and protect it from unauthorized use. As noted earlier Starbucks and others that have stored data for their easy pay system in simple unencrypted text have been charged with being less vigilant in their data protection.

Yet despite the claim by Apple Pay that it is a highly protected service, there have been problems since this system was launched in October 2014. As reported on March 24, 2015, in the *Washington Post* and other media: "There has been a sharp rise in reports of fraudulent Apple Pay transactions." This has led to accusations that this easy pay system is "too easily subject to credit card fraud." Apple has fired back to say that the problem is not with the Apple Pay system but with the credit card banks authentication system. Apple has explained that its unique identifier code cannot be "hacked" and that it is the rapid authentication process that credit card banks use that grants access to the Apple Pay system in only a few seconds that needs to be re-geared. In short, the Apple Pay experts claims that there is a need at the outset to scrutinize more closely if the credit card number may have been stolen or that there are other credit issues that would stop the instantaneous entering of the card into the system. One payments expert, Cherian Abraham, has suggested at as many as 6 % of all Apple Pay transactions have involved stolen or compromised credit cards. If this were true it would be a rate that is 60 times greater than the industry average of 0.1 %.

The consistent reports of fraud and "phantom charges" is currently giving retailers and banks some pause. Apparently the decision at Apple as to whether to register an Apple Pay user was as simple as looking at i-Tunes credit reports, and the banks, in the rush to join in on the Apple Pay bonanza, did not use the same rigor that is used with credit card applications. The greatest concern currently seems to be in the move of Apple Pay from brick and mortar shops to online shopping, where cyber-crooks with stolen credit card numbers might be able to increase their rate of transactions. As Amitabh Saxena, founder of Digital Disruptions, has said: "I think there is a legitimate concern for online merchants. Clearly there needs to be more clarity as to where the breakdown in the security of Apple Pay transactions are taking place, and credit card banks need a more demanding process at the front end of allowing consumers into the system" [6].

In theory a credit or debit card using near field communications theoretically cannot be read unless it is within 10 cm (4 in.) of a reader. But, there are creditable reports of transactions being recorded as far away as 1 m (just over 1 yard). Fortunately, if your card is stolen the thief can never spend more than about $30 at a time. If the contactless payment is used several times in a short interval it produces a PIN request. Further, credit card banks are required to cover your losses as long as you have acted with "reasonable care."

Nevertheless cyber experts have demonstrated that the NFC codes stored on mobile phones nonetheless can be intercepted by enterprising cyber criminals. Kits that can be purchased for under $100 are actually able to receive the very weak signals from NFC contactless cards or phones. The NFC reading kit scans the details from a card for future use. If the cyber thief knows to space out purchases or cash requests against debit cards below the limit, it may be some time before the cyber-fraud is detected [7] (Fig. 5.2).

The technology of NFC cards and mobile phone applications is simply an extension of so-called radio frequency ID (RFID) technology that was first used for industry inventory control some 15 years ago. RFID represents as a step beyond computer bar codes. The technology of near-field communication uses electromagnetic induction between two loop antennas located within each other's near field. The frequency that is used for this purpose is radio frequency in the industrial, scientific and manufacturing (ISM) band of 13.56 MHz. There is also an adopted standard air interface for NFC that is designated by the International Standards Organization ISO/IEC 18000-3 that allows for data rate exchange at rates ranging 106–424 kbit/s. The tags that can be read by an NFC reader can not only store personal data such as

**Fig. 5.2** An NFC-enabled mobile phone interacting with a smart poster

debit and credit card information, but it can also provide loyalty program data, among other such information. One of the more unusual applications is to allow the life story of a deceased to be stored on the tombstone so that it could be read by someone with the right app on their cell phone.

The burgeoning growth of Apple Pay, Oyster, Maestro PayPass/Visa PayWave/blink/American Express's Express, Pay Pal's "Paydiant", etc., is making NFC cards for touch and go payments a worldwide phenomenon.

Visa and MasterCard, in response to security concerns related to contactless payments, are providing protection plans. MasterCard offers Zero Liability protection against unauthorized purchases. Visa also has its own Zero Liability policy and says its contactless cards are "as secure as any other Visa chip card." The key to the contactless credit card is the fact that it has its own unique 'key.' Each time a transaction takes place, that key uses encryption technology to generate a unique verification value or authentication code. When the transaction takes place, the issuer checks that code before deciding to accept or reject the transaction.

One form of protection that is a relatively low cost and efficient form of risk reduction is to buy a RFID-shielding wallet, pocket, or credit card sleeve that is designed to shield your card. There are products such as Blackout Pocket that is designed for this purpose. But if your tap-and-go app is on your mobile phone, it is much harder to shield in a convenient and efficient manner [8].

# The Cloud

Your first reaction to this topic may well be that you don't need to use "The Cloud." So the security of Cloud-based services is not a worry for you. You don't store massive amounts of data or access super computers to obtain massive processing power. No one in your household is attempting to decode genomes or simulate what happens inside an atomic collision in a nuclear accelerator at the Fermi Research Lab, so why is this relevant to you?

Actually there are two answers. Individuals do want and need to store a good deal of visual information and graphics and sometimes their Internet service provider limits the amount of information that they can send and receive online. A low-cost back up of all your data is a good way to shield your computer against a cyber-attack.

Thus it really is possible for individual users to sign up for a very modest monthly fee to access a large amount of visual data backup. This also allows you to store large files and high definition images on a sort of "drop box" that you can share with others that you choose to designate. This is indeed a service for individual consumers and can come at a cost as low as for free.

The second reason is that many of the companies and institutions that you rely on daily are actually significant customers of Cloud storage, software and processing capabilities. The town, city, or county in which you live probably stores information about your taxes, your real estate holdings, and perhaps your marriage license or court records on The Cloud. Your banker or stock broker may have all of its vital financial records and information on your stock or bond holding stored on The Cloud. Governments around the world have increasingly moved to storage of records related to taxes, tax audits, Social Security or federal health insurance services on The Cloud. This is because the storage rates are very low, and there is reliable offsite back-up storage to protect key data that is maintained in this manner. A hurricane, tornado, flood, fire, or other type of disaster could threaten vital data unless it is stored in a protected site in another location.

## *Evaluating Personal Cloud Services (Fig. 5.3)*

Guide to Personal Cloud Services and Their Cost and Performance

There are many options online now available for personal Cloud services. Users in choosing the service for them would perhaps wish to consider the available options in terms of speed, reliability, security, ease of use, cost, and

**Fig. 5.3** The ever more omnipresent world of Cloud computing

their level of service support. Most consumers will likely consider cost, reliability and security to be the most important factors in choosing the best Cloud storage provider. Have a look at some of the highly rated Cloud storage services below [9] (Table 5.2).

Business Cloud Services

Cloud computing lets you rent the tech you want, and most typically you will be paying a monthly fee for the service that you opt for. These Cloud services typically break down into three categories. As the chart below indicates one can choose to obtain SaaS (Software as a Service) or PaaS (Platform as a Service). PaaS means renting everything, but the app (often used by app developers is especially for mobile apps—i.e., smart phones), while Infrastructure as a Service (IaaS) means renting just the hardware and tools to maintain the hardware (popular with startups and enterprises) [9] (Fig. 5.4).

The assumption that many would make is that the organizations that are offering Cloud services are very sophisticated and thus use of The Cloud is very secure. The more complicated answer is that its use can be quite secure and that it can offer a multitude of services, but nevertheless care has to be

Table 5.2  The top three personal cloud computing backup services

---

*Just Cloud*

This service has recently become one of the more popular Cloud storage providers. Its unconventional strategy was to begin by offering, free, no obligation online storage space and in an unlimited amount. Check the website for current prices and promotional offerings. Just Cloud is also known for fast upload/download service. This is due to an infrastructure that incorporates Google "Big Data" storage and Amazon s3 to maximize speed. Just Cloud has made cloud storage simple by automating the whole upload process. Its software automatically backs up all your files to your free Cloud storage space so you can access your files anywhere at any time. New files and updates are automatically uploaded.

---

*ZipCloud*

This service is currently available at a cost of $4.95 per month. It places emphasis on security through encryption. It offers those users that are more computer proficient a large number of useful features. Zip Cloud is a very high speed and user-friendly product that collects computer data and compresses in order to be transferred to a secure Cloud. A highly secure 256-bit AES encryption shields the data to enhance its security for the personal computer user who puts a premium on risk reduction. It is limited to 250 gigabits of storage, which is still sufficient for most users.

---

*MyPCBackup*

This site is available at a cost of $2.95 a month. It also makes online computer backups by means of a simple and automatic backup function with fast upload speeds and convenient features. MyPCBackup offers plenty of online storage space, an easy-to-navigate members area, and an intuitive desktop application that allow you to access your files from anywhere at any time. You also have the ability to sync multiple computers to one account and unlimited storage.

---

*Others*

There are many, many other providers of this type of backup, large storage, encryption, and other services with differing features and prices. Other highly rated providers include: Box, Dropbox, Google Drive, Apple iCloud, Microsoft OneDrive, SOS, Carbonite, Sugar Sync, BackBlaze, Spideroak, Hightail, and Livedrive.

The costs by the various providers for storage, editing, and other services can vary from $3 a month to $16 a month, and features vary. The description of the service and the price is spelled out on their various websites.

Some of these providers, such as Box, Drop Box, Google Drive, Apple iCloud, and Microsoft OneDrive provide a free low volume storage and file-sharing capability, but then offer a premium service as a byproduct of a richer and more robust service menu that includes a significantly larger file storage size that ranges from 100 gigabytes to even 1 terabyte of storage. There are lots of charts that show different vendors in the "top ten." A review of all features from vendor websites is recommended before making a final decision. The bottom line is that there are many providers that offer quite good services below $10 a month and many that are lower than $5 a month.

---

taken. Businesses or governmental agencies that use such services need to employ encrypted networking and also review in detail the precautions taken by their Cloud provider. Figure 5.5 shows a breakdown of how services can be moved from everything in-house (all blue) to progressively moving responsibilities for computer services essentially to out-of house (gray) with everything being in the Cloud. As one moves from IaaS to PaaS to SaaS, of course, the fees for use of Cloud services increase to an ever higher monthly fee [10].

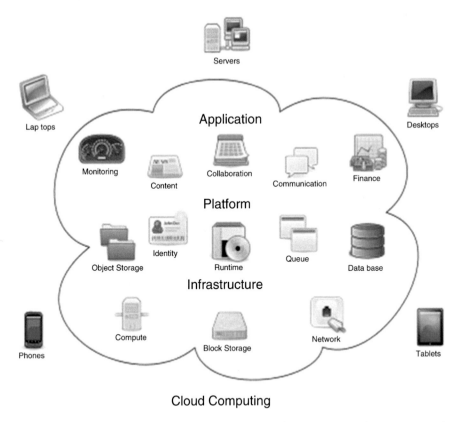

**Cloud Computing**

**Fig. 5.4** Cloud computing offers multiple ways to obtain access and many types of services

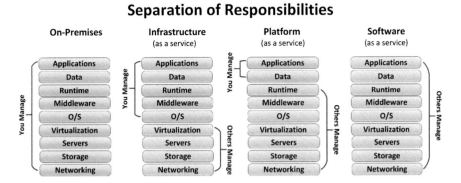

**Fig. 5.5** A chart showing how computer services can be shifted to The Cloud

Google released a report in early April 2014 about flaws in a system known as Heartbleed in the form of a major online security flaw. This flaw allows access to Open Secure Sockets Layer (Open SSL) web servers. Heartbleed allows cyber criminals to access website data and visitors' personal information, including credit cards, e-mails, and passwords that are stored in The Cloud. Unfortunately Heartbleed leaves no record in an attacked web server's logs, which makes it impossible to tell exactly how many websites may have been exploited by it. Heartbleed went undetected for more than 2 years, and it could have affected thousands of OpenSSL web servers across the globe. The U. S. Department of Homeland Security has now warned businesses about Heartbleed and asked them to review their web servers to find out if they are using infected versions of OpenSSL. This may well be the biggest single security breech that individuals need to worry about in terms of their own personal financial security [11].

Unfortunately Cloud issues go beyond security lapses to Heartbleed and supposedly hack-proof OpenSSL. Usually, when one goes to an HTTPS site, there is a reassuring logo showing a little lock that is supposed to give you confidence that the SSL functionality within the Internet transport security layer is protecting your transmissions and online activities from the snooping eyes of black hat hackers. Remember that HTTP sites are NOT protected, and even the HTTPS sites have their security flaws.

Five Kinds of Cybersecurity Concerns with The Cloud

John Kinsella, the founder of Protected Industries, has outlined five key security issues associated with the use of The Cloud. Although Kinsella's analysis was initially presented in 2012, the issues remain today. Any chief technology officer or data administrator charged with overseeing the protection of data that is stored on The Cloud or using applications stored there should consider the following Cloud security issues. The following is a paraphrase in simplified terms of the vulnerabilities that Kinsella has flagged for attention [12].

1. INTERNAL CLOUDS, ALTHOUGH PRESUMABLY MORE SECURE THAN PUBLIC CLOUDS, ARE NOT INHERENTLY SECURE, AND THUS SECURITY BREACHES ARE POSSIBLE IN A VARIETY OF WAYS.
   It has been assumed by some chief technology officers that if the commercial public Clouds available are not entirely secure, that the answer is to construct a private Cloud that can be secured behind a firewall. This would seem to be a logical solution for an organization with a good deal

of intellectual property that must be totally secure. As Kinsella has noted this is not a failsafe solution. He has noted that it is still possible for a security breach to occur: "It still takes just one bad apple to spoil the barrel—a single department, user or application that is not behaving as it should." Just as a single individual accessing a secure corporate site via an unprotected wireless LAN or router can disclose key corporate data, it takes only one user of a Cloud that is careless with access passwords or critical data for a security breach to occur [9].

2. COMPANIES OR ORGANIZATIONS THAT USE A PUBLIC CLOUD WILL OF NECESSITY HAVE A LESSER DEGREE OF SECURITY VISIBILITY AND MONITORING CAPABILITY. IN USING A PUBLIC CLOUD THERE WILL BE LESS RISK AWARENESS.

There are a number of strategies that can still be used to alert users of The Cloud of possible security issues even when using a public Cloud. Such systems' controls and monitoring capabilities could include the following strategies: Log more information into your applications and set up systems to generate alerts when signs of compromise or some sort of cyber-attack is detected. Alerts could come in such cases as when a file is modified, records are changed more frequently than usual, or resource usage is abnormally high at a particular time.

This type of monitoring and detection of attack strategy would not work in the case of software as a service (SaaS). The bottom line is that when an organization decides to move to The Cloud it should be clear that there is de facto going to be a lack of visibility, and where possible there needs to be mitigating controls to adapt to the new conditions. If the intent is to provide software as a service (SaaS), the nature of the mitigating controls will need to be more sophisticated and probably less effective. Such alerts should be possible in the case where the service is either Infrastructure as a Services (IaaS) or Platform as a Service (PaaS).

3. YOUR MOST SENSITIVE INFORMATION NEEDS SAFER STORAGE. THIS MOST PROBABLY MEANS A SOPHISTICATED APPROACH TO ENCRYPTION. IT COULD ALSO MEAN THAT SOME OF THIS INFORMATION MIGHT NOT BE STORED OR PROCESSED IN THE CLOUD AT ALL.

This is merely common sense. Protect your most valuable assets the most securely. A wise person once said: "If you protect your diamonds and your office supplies with the same degree of diligence, you will lose more diamonds and fewer office supplies."

If you have particularly sensitive information that is in need of special protection then take steps to do so. One approach is, of course, in the encryption of the data or your most valuable intellectual property.

Kinsella particularly advices that one does not store the encryption key with the encrypted data. If you provide the key with encrypted data you have defeated the purpose of the encryption. Several companies make appliances (both software driven, i.e., "virtual" or physical hardware) that proxy data leaving an office. In short data, and particularly critical data, is encrypted or "tokenized" before sending it to The Cloud. Thus these users access The Cloud service from behind that appliance, which provides them security protection if they consistently use this process before any data is sent to The Cloud for storage or processing.

4. COMPUTER APPLICATIONS AREN'T NECESSARILY SECURE.

As discussed in the mobile section of this chapter, applications are not secure and can be vulnerable in lots of ways. This starts with the fact that the application may have been developed by tools or kits that are vulnerable to attack. Apple claims their apps are more secure because all tools or kits that are used for their apps depend on their own proprietary tools. The important "apps security" increases when particular applications are used in a Cloud environment. This is simply because the application is exposed to more potential accessing users that are seeking a security vulnerability. If possible the application might be redesigned for deployment in The Cloud to make it more resilient against a black hat hacker or corporate competitor.

5. AUTHENTICATION AND AUTHORIZATION MUST BE MORE RESILIENT AND DILIGENT. THIS IS PARTICULARLY SO WHEN USING THE CLOUD FOR A PLATFORM AS A SERVICE, INFRASTRUCTURE, AND FOR STORAGE.

There are many Cloud authentication and authorization tools and solutions available. In fact, this task has the greatest number of commercial solutions available. This does not mean that there are not concerns here to consider. Questions to answer include such issues as: Is it flexible enough to accommodate new services? Does the monitoring system provide alerts to a series of authentication attempts that are rejected in short succession?

This is not to suggest that the above cybersecurity concerns that are reviewed above represent the only concerns that users of The Cloud should consider, but these do represent perhaps the most prevalent problems and the ones that CIOs and IT security managers should consider at the top of their checklist of ongoing concerns. As more and more computer networking activities in terms of application development, software usage, processing power, platform access, encryption of vital data and intellectual property and storage of information flows to public and private Clouds vulnerability to hacker attacks of various kinds will likely increase rather than decrease.

The nature of those risks and vulnerabilities should not be looked at solely in terms of cyber-attacks by cyber criminals, corporate competitors, and techno-terrorists. It turns out that old-fashioned problems such as power outages, loss of signal, and even what is sometimes called "back hoe fade" (i.e., when a tractor or digging machine cuts through a vital coaxial cable or fiber network connection) could shut down your operations for an hour or even a day or a week. In short, when you are looking to possible high-end ways to increase your cybersecurity in the use of The Cloud, such as encryption and appliances that can "tokenize" your data before it goes out of your shop, don't forget the basics. Remember that power outages and breaks in cables can shut you down or result in data loss and communications breakdown just as surely as cyber-criminals on the prowl.

The landscape of who is providing Cloud services continues to shift. Amazon World Services is currently the largest, but Microsoft is currently gaining ground, with both have close to $6 billion in annual billings. Since the market size continues to grow and billings continue to rapidly increase, it is perhaps most useful to show what types of services the largest providers offer and to note that they are significantly different as shown in Fig. 5.6, as contained in the Gartner Group analysis [13].

## Summary of Major Vendor Emphasis

| | Build Private Services | Deliver Services | Services Delivered* | | | Private Offerings Enabling Tech. | Packaged Cloud |
|---|---|---|---|---|---|---|---|
| | | | IaaS | PaaS | SaaS | | |
| **Amazon** | ○ | ● | ● | ◑ | ○ | ├— None —┤ | |
| **salesforce.com** | ○ | ● | ○ | ● | ● | ├— None —┤ | |
| **Google** | ○ | ● | ◑ | ◐ | ● | ├— None —┤ | |
| **Microsoft** | ● | ● | ◑ | ● | ● | ├——●——┤ | |
| **IBM** | ● | ◑ | ◐ | ◑ | ◑ | ├———●——┤ | |
| **VMware** | ● | ◑ | ○ | ◐ | ◑ | ├———●—┤ | |
| **Oracle** | ● | ◑ | ◑ | ◑ | ◑ | ├——●——┤ | |
| **SAP** | ○ | ◑ | ○ | ◑ | ◑ | ├— None —┤ | |
| **HP** | ● | ◑ | ◐ | ◑ | ◑ | ├————●┤ | |

Note: This is not an evaluation of capabilities, but rather of emphasis.

● ◐ ◑ ◒ ○
Significant ——————→ None

\* The provider may offer public, community or virtual private services

**Gartner.**

**Fig. 5.6** Areas of emphasis for various providers of Cloud services. (Graphic and analysis courtesy of the Gartner group.)

# Conclusions

In a survey of top corporate information officers (CIOs) as to what keeps them awake at night with regard to computer network security, it turns out that mobile and wireless access to corporate data and networks scored number one, and access to and using The Cloud for computing services that had not been adequately secured scored number two. The advantages that either private or public Cloud technology can add are substantial. Cloud providers offer flexibility, cost savings, and access to a variety of powerful tools. The Cloud can allow application developers to have the same intellectual resources that large corporations have at their disposal. But as in all things, there are catches.

For the ordinary user, it is important to know that mobile communications, Wi-Fi, and wireless LANS offer a variety of risks and that you need to obtain and keep up-to-date protection not only for your desktop computer but for your smart phone as well. There are also cautions that you should note about instant credit card payment systems, and to know that your information could be captured by someone trying to steal your credit card information. Use of The Cloud for your personal computer backup and for file sharing with others is a useful thing to do. This carries little risk if you do not include vital information in the files that you have backed up in this matter. Storing your passwords offline is a very good idea. We hope the advice provided in this chapter can prove helpful in terms of how you protect your smart phone, use wireless networks with less risk, and also learn that Cloud computing can be a useful capability, but it, too, harbors security risks.

# References

1. Avast Global Hacking Experiment Shows How Exposed Wi-Fi Users are https://blog.avast.com/2015/03/02/avast-study-exposes-global-wi-fi-browsingactivity/?p_pro=1&p_vep=10&p_elm=55.
2. Katherine Muniz, "Banking Apps on Android Phones Are Vulnerable to Risk" February 16, 2014. http://www.fool.com/investing/general/2014/02/16/banking-apps-on-android-phones-are-vulnerable-to-r.aspx.
3. *Ibid.*
4. Sid Kirchheimer *Scam-Proof Your Life: 377 Smart Ways to Protect You & Your Family from Ripoffs, Bogus Deals & Other Consumer Headaches* (2007) Sterling Press, New York.
5. Gerald Morales, "3 Popular Apps That Could Put Your Money At Risk" *The Motley Fool,* February 9, 2014.

6. Hayley Tsukayama and Sarah Halzach, "Fraud raises question: Is Apple Pay Just Too Easy?" *Washington Post*, March 24, 2015.

7. http://www.theweek.co.uk/prosper/53317/contactless-cards-what-are-risks#ixzz3VErxStmW.

8. http://www.creditcardoffers.com.au/guide/2013/12/tap-go-really-encourage-criminals/.

9. "Top Ten Best on Line Backup," http://www.thetop10bestonlinebackup.com/cloud-storage.

10. http://www.businessinsider.com/10-most-important-in-cloud-computing-2013-4?op=1#ixzz3VjkrKDbh.

11. Dan Kobialka, "Heartbleed Open SSL Security Flaw Puts Corporate Cloud Data at Risk" *Talkin Cloud*. Apr 9, 2014. http://talkincloud.com/cloud-computing-security/040914/heartbleed-openssl-security-flaw-puts-corporate-cloud-data-risk

12. John Kinsella, Protected Industries, "5 (More) Key Cloud Security Issues" Sep 26, 2012. http://www.csoonline.com/article/2132313/cloud-security/5 – more – key-cloud-security-issues.html.

13. Louis Columbus, "Demystifying Cloud Vendors" Forbes, Feb. 20, 2013. http://b-i.forbesimg.com/louiscolumbus/files/2013/03/Summary-Chart.jpg.

# 6

# Protecting Vital Cyber Infrastructure

## Introduction

The world of Internet security unfortunately continues to become more and more complex. Two of the areas of cybersecurity concerns that are among the most technically intricate involve satellite communications and remotely controlled and automated IT systems called Supervisory Control and Data Acquisition (SCADA) systems. This chapter addresses these two areas because they unfortunately can expose us to lots of cyber dangers.

We indicated at the outset that this book would try to avoid techno-speak and gobbledygook, but we have to admit that this chapter may a bit more complex than the others. We suggest that if parts seem a bit technically turgid, just skip over them and leave the solution to technical experts in government and industry that are trying to fix the problems and issues we discuss here. The good news is the rest of the book remains quite straightforward and that if you choose to skip this chapter you can do so with a clear conscience.

Unfortunately cybercrime and the number of its practitioners continue to grow along with the complexity and exponentially increasing patterns of use. During a typical week in the United States there are well over a half million hacker attacks. The number of cyber incidents involving U. S. government agencies jumped 35 % between 2010 and 2013, i.e., from roughly 34,000 to about 46,000, according to another recent report by the Government Accountability Office. In 2014 the rate has sharply increased with almost 61,000 cyber-attacks and security breaches across the entire federal government for the calendar year of 2014 according to the official report from the White House [1].

© Springer International Publishing Switzerland 2015

J.N. Pelton, I.B. Singh, *Digital Defense*, DOI 10.1007/978-3-319-19953-5_6

One needs to put these cyber-attacks into some perspective. During surveys taken in March 2015, when the U. S. government was collecting data on total systems use, it found that there were on average over 1.5 billion consumer interactions with American governmental websites per week, or somewhere around 75 billion interactions a year. The good news is that if one divides 75 billion total interactions by about 60,000 cyber-attacks, this translates into only 1 attack in every 1,250,000 interactions. The bad news is that the ratio of attacks continues to increase.

There are also ongoing concerns as to whether all federal agencies have adopted the mandated Federal Information Security Management Act of 2002 for the protection of federal data against cyber-attacks and maintain privacy of information consistent with the guidelines established by the National Institute of Standards and Technology (NIST). In a recent audit carried out by the General Accounting Office, the Library of Congress's Chief Information Security Office listed 30 systems that were subject to security oversight, but the library's Information Technology Office initially reported this as being 46 in number, and after further audits it was revised to be 70 in number. It is far from clear that the Library of Congress example is indicative, but if so the problems of cyber-attacks and privacy incursions may be much greater than reported [2].

As we will learn in the next chapter, the challenges will only become greater as the future unfolds. The fact that billions of people are using mobile phones to connect to the Internet and billions more are connected via desktop computers and wireless LANS typically places the greatest focus on cybersecurity with these user connections.

Yet there are the other cyber concerns that are important to be aware of and to know what protective strategies are available to respond to their various other types of cyber weaknesses. There are two cybersecurity concerns that are largely hidden from view. One is hidden in plain sight and close at hand, even as close as a light switch in your bedroom, the tap water in your kitchen, or the traffic signal in your own neighborhood in your town. When you turn on your water, flick on your lights, turn on your gas oven, flush your toilet, buy gas at the service station, or encounter a traffic signal, there are automated controls involved that, if tampered with, could cause you and your community harm. These controls are known as SCADA. This obscure technical term, as we've discussed earlier, stands for Supervisory Control and Data Acquisition. This is the formal name for computerized wireless network control systems that remotely regulate oil, natural gas, water, and sewer pipeline flows. SCADA networks also control the electrical grids in your town or city, and regulate the timing of traffic lights. If you saw the movie *The Italian Job* (2003) with Mark Wahlberg, Ed Norton and Charlize

Theron, the way that the traffic signals were hijacked to steer the armored car so it could be robbed was by taking control of the SCADA network. If cyber criminals or even worse cyber-terrorists were able to take over a SCADA network they could do bad things. And the scary thing is that SCADA security is only recently being seen as a major risk factor and one that should be given priority attention.

The other unseen aspect of cybersecurity is as far away as SCADA systems are near at hand. This is the amazingly complex and important network of satellite telecommunications that fly high above Earth's surface hidden from view. Few people realize how important satellites are in their daily lives. People in rural and remote places, hundreds of millions of them, now depend on communications satellites not only for television and radio entertainment but for communications, data networking, as well as for weather forecasting, positioning and navigation, and even timing. Did you know that GPS satellites provide the timing for global synchronization of the Internet. If we lost these satellites due to a natural disaster or a cyber-attack, we would lose the ability of the Internet to operate globally in the course of a day unless back up timing systems were available.

Thus in this chapter we will first explore how we depend on SCADA systems and what we can do to protect them. Then we will examine the fascinating world of application satellites, how we depend on them, and what we can do to protect them from attack or interception by cyber criminals and techno-terrorists.

## SCADA Networks in Our Lives that We Never See

Since SCADA systems are essentially remote control systems that run virtually every utility and infrastructure in our modern world it is essential that they operate continuously and reliably. These systems control rail switches, electrical grids, all sorts of pipelines, and traffic lights. They have become our pervasive servants, carrying out billions of orders a day through computer networks. Without very many people recognizing it SCADA networks have become essential to sustaining our modern way of life and are almost everywhere in our daily lives.

One may think that the key is to create secure and encrypted SCADA networks with protected passwords of some complexity that are regularly updated and with an audit function to ensure this level of security. And they would be right in thinking this level of security needs to be developed and implemented. Unfortunately communities and utility and transportation companies are only slowly making this so. But there is another concern as well.

These SCADA systems can also be wiped out or degraded in their performance by natural hazards. Issues that we have to be concerned about include solar storms, earthquakes, floods, and fires. All of these natural hazards and more could impact SCADA performance, directly or indirectly. If a natural hazard wipes out a power supply then SCADA systems, if not backed up by uninterrupted power supplies and generators, can be taken offline as well. In short it is essential to protect the resilience of SCADA systems against a vast array of threats, such as from cyber criminals, techno-terrorists, natural disasters, and widespread power loss. In light of the complexity, pervasiveness, and interconnectivity of these SCADA systems this is hard to do.

For example, a blackout in power grids will cause electricity distribution problems within the SCADA system and so result in traffic light control problems and railway shut downs, causing traffic chaos. Water and sewage systems and pipeline operations would likewise shut down or malfunction. Thus, there is the need for taking a holistic and strategic approach in defending SCADA systems from fires, earthquakes, hurricanes, extreme solar events, man-made EMPs, and direct cyber-attacks. We must consider various elements that contribute to the operation of SCADA systems. These include not only defensive technological solutions, but key processes, polices and their compliance; and finally people awareness and training.

Many SCADA networks operate within a corporate complex such as an oil field or a mining operation to control field devices, as shown in Fig. 7.2 below. Although these operations are not as critical as a SCADA system that controls the cooling operations at a nuclear power plant or the switches on a rail line, these could still be used by a cyber-terrorist to create a major industrial accident with a large-scale loss of life. Those who are in charge of SCADA networks need to consider the worst possible accident or cyber-attack that might happen or be caused to happen and work to prevent it (Fig. 6.1).

SCADA owners and operators should seek to develop protection programs against industrial accidents, network shutdowns, or system failures as follows [3]:

1. Put in place a system architectural design for resilient capabilities and communications strategies that allows for redundancy, backup controls, and emergency overrides.

2. Insure that back-up power supplies such as an Un-interruptible Power Supply (UPS) and generator are in place and periodically checked for operation.

3. Implement integrated back-up timing systems to accommodate the temporary loss of GPS because of interference or failure.

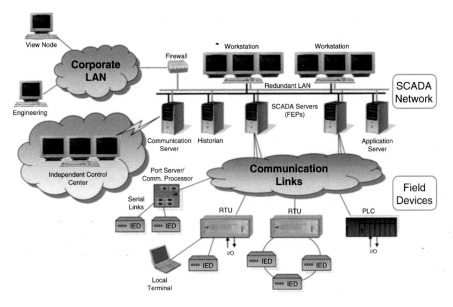

**Fig. 6.1** A typical corporate-based SCADA system that remotely controls field devices

4. Absolutely vital systems should be shielded and protected as much as economically viable. There are other options, such as to employ encrypted machine-to-machine communications networks that could back up or perhaps even replace SCADA systems. These would need back-up power capabilities and other precautionary steps taken as well to protect those systems that need failsafe capabilities.
5. Owners and operators of SCADA systems that are not essential for real-time system operations should consider switching off key electronics or disconnecting the entire SCADA system during high alert periods.

## SCADA Systems Security in a Community

As chair of the IT commission in a local community in the Washington, D. C., area, the question that came before our commission was how safe and secure were the SCADA networks against cyber-attacks. The answer that came back on investigation was "not very secure at all." There were SCADA networks for traffic control under the Transportation Department. There were separate SCADA networks for water and sewage pipelines under the utilities department. There were yet other networks under the control of other departments. These department were focused on doing their job and not cybersecurity. Many of the SCADA networks had "back doors" that had the original codes from the manufacturers of the equipment and had not been changed in years. No one was charged with auditing the security of these SCADA networks.

Fortunately this was not a difficult problem to fix once an outside consultant had reviewed the security of the community's SCADA networks. The access codes were all changed to passwords that were rigorous in nature. The security of all the various systems were moved to the Department of Technology Services that had computer expertise. A periodic audit of the SCADA network and other broadband IT networks for the government was instituted to insure that a techno-terrorist could not hijack that community's SCADA system and send a command to open a sewage valve and contaminate the entire community's drinking water supply or turn all of the traffic signals green. But this was only one community that had one of the highest educational and income levels in the United States. There are many other communities in the United States and around the world where SCADA networks are largely unsecured. You might be well advised to ask your local government as to what they have done to secure the community's SCADA networks, the broadband telecommunications and IT networks for the local government and the schools, and other public or private networks that might be attacked by cyber criminals or terrorists [4].

## Satellite Networks and Cyber Risks

There are many Internet applications that are now downloaded via direct broadcast satellite systems. It makes far more sense to download new software to black boxes that control television programming and Internet access rather than replace millions of black boxes.

Further there are millions of consumers in remote areas not served by fiber optic cables or businesses not adequately served by broadband terrestrial systems that rely on communications satellites for asymmetrical Internet services. And there are concerns with Internet security that go beyond broadband satellite telecommunications service, and involve satellite positioning, navigation and timing satellites—especially the GPS network that plays a critical role in the time synchronization for the global Internet.

Then there is the entirely different set of issues that relate to the ubiquitous control systems that are core facilities for operating vital facilities in transportation switching and routing, electrical power systems, pipelines, water and sewage systems, traffic control signals, and much more. Millions of these SCADA systems play a critical role in the automated control of critical infrastructure in the United States and across much of the world today.

## Satellites and Cybersecurity

Millions of people in the United States and around the world live in large cities where fiber optic networks provide broadband connectivity to homes and business. In the location where the authors live there are two competitive commercial broadband cable television, telephone, and broadband service providers. In addition their local community government is installing a fiber optic network to connect 100 governmental and school sites with an additional installation of 686 dark fiber units that will be made available to businesses and governmental agencies for truly broadband services up to a gigabit per second. But the availability of such tremendous throughput capability is simply not available in many parts of the United States, Europe, OECD countries, and especially in rural and remote parts of the world.

There are increasingly sophisticated satellite networks that are being designed to support broadband networking and telecommunications services around the world. There are actually a variety of satellite designs and new types of networking and ground user terminals that are being deployed to meet the demand for broadband services in areas where terrestrial systems cannot provide this service.

Coaxial cables and fiber optic networks can be attacked by cutting them apart or using sophisticated technology such as devices known as "limpets", but a cyber-attack on them is much more difficult than wireless systems. Just as one can attack smart phones, wireless LANS and Wi-Fi systems it is possible to eavesdrop on satellites and a newly developing technology known as High Altitude Platform Systems.

For years the idea of having a floating device or stationary aircraft serve as an observation post or platform for telecommunications and other services has been the subject of experiments and speculation. So-called aerostats have been used in various locations around the world with varying success. The International Telecommunication Union (ITU) has now allocated frequencies (12 GHz uplinks) and (14 GHz downlinks) for this type of telecommunications service. Robotic aircraft and more sophisticated airships are now being designed for this type of service. Such a high-altitude platform system (HAPS), operating at 21 km (14 miles) high. could provide

**Fig. 6.2** An experimental airship conceived to support high altitude platform system (HAPS) services

telecommunications services for a Cuba, a Jamaica, or other small country. Because of the small amount of "path loss" associated with transmissions over such short distances and the possible use of advanced cellular technology a HAPS could support quite a large broadband service at reasonable cost. This would be a sort of high-flying Wi-Fi system [5] (Fig. 6.2).

It is not certain as to how soon we will see a significant number of HAPS for telecommunications and broadband Internet services nor whether these will be airships or robotic aerial vehicles or some other technology, but essentially all of the cautions and concerns provided with regard to Wi-Fi systems and mobile apps used via smart phones to connect via wireless networks apply. Just as we have seen cybercrime follow Internet innovations into new fields, it can be predicted with certainty that as HAPS are deployed cybercriminals will exploit this technology as well.

## Satellite Technology and Systems

To a very large extent our modern lives and livelihoods are controlled by technologies that have become invisible to us. Electrical engines, computer processors (not only in computers but in virtually everything else from cars to washing machine to toasters), all sorts of utility systems, traffic lights, and now satellites, yes satellites, are a part of our daily routine. We get up and check the weather, but the weather doesn't come from the weathercaster, it comes from weather satellites. We listen to the news from around the world. Your home or business may be served by cable television, but most of the news originates from satellite feeds uplinked from around the world. There are today around 20,000 television channels operating around the world. Global events such as the Olympics and the World Cup games depend on scores of satellite feeds.

However, it doesn't stop there—not by a long shot. You get online to check out your e-mail and never suspect that the precise timing that allows the Internet to function is controlled by the atomic clocks in the GPS positioning navigation and timing satellites. Nearly half the countries in the world depend on satellites to connect to the Internet. You get in your car to drive to a new location, and your vehicle is guided by navigation satellites. You know you pay taxes to fund the military forces to defend your country, but you never think that so many of the military and defense operations are depending on weather, navigation and guidance and telecommunications satellites.

There are several scary videos on line that are entitled: "A Day Without Satellites," and the story that is told is unsettling. Banking, commerce, transportation, military and governmental operations, the Internet and communications and radio and television broadcasting, and health and education services all shut down or are very strongly impacted. We live in a world that is far more dependent on these servants in the sky that we ever knew or suspected. (Incidentally the name "satellites" was coined by Galileo when he first saw the moons of Jupiter scurrying around the giant planet acting like servants to their master planet. The ancient Latin word *satelles* actually means servants.

This book is not about satellites but about cybersecurity, but the world of satellites and cybersecurity are closely related in several ways. Cyber criminals could use satellites to spy on you. In fact, listening in on satellites is one of the ways that the NSA and other intelligence agencies around the world collect information on terrorists, drug dealers, and people contemplating military attacks or other criminal acts. When news reports talk about "chatter" they are talking about monitoring of Internet and telephone and data exchanges over all sorts of media from cell phones to satellites and sanctioned eavesdropping over every form of electronic media.

In order to explain the cybersecurity issues that involve satellites a little bit of a primer will be provided about different types of satellites, what they do, how they might be attacked, and how cyber criminals or techno-terrorists might use these devices to commit crimes or engage in terrorist attacks. On the other hand satellites can be a useful tool to track down cyber criminals and terrorists and provide for a military defense as part of a national or regional defense system.

## Application Satellites with Cybersecurity Implications

We will start with the simplest issues first. There are three types of application satellites that can have some impact on cybersecurity. These three types of satellite systems are: (1) weather or meteorological satellites; (2) Earth observation, remote sensing, or surveillance satellites; and (3) what are officially known as Positioning, Navigation and Timing Satellites, but most people simply know them as the Global Positioning Satellite (GPS) system, which is the system that the United States operates. Other countries such as Russia, China, Japan, India, and Europe have their own types of these satellites, but the United States tends to be leader in many of these type satellite systems. There is a fourth type satellite system, namely communications satellites, but these are the most relevant and complicated, so we will address these satellite networks last [6].

The concerns here largely center on techno-terrorists who might attack these satellites or even solar storms that could disable vital satellite operations. There could be ways, however, that cyber criminals could attack using our servants in the skies against us.

WEATHER OR METEOROLOGICAL SATELLITES. National defense systems rely on weather satellites for their operation and have separate meteorological systems on the ground and in space to monitor vital storms as well as local weather conditions where hostilities are taking place. There are even satellite systems to monitor the Sun, to detect and defend against solar flares (electromagnetic phenomena) and even more dangerous coronal mass ejections (streams of ions ejected from the Sun at millions of kmph speeds) that can wipe out electrical power grids and yes, various application satellites. In short these weather satellites are a part of a national defense intelligence system, and a cyber-terrorist that wished to attack a country might conceivably try to get command codes that could be used to disable one or more of these satellites or send it into a new orbital path, where it might crash into another satellite to destroy or disable both [7].

EARTH OBSERVATION, REMOTE SENSING AND SURVEILLANCE SATELLITES. A weather satellite is actually a specialized form of Earth observation satellite.

Satellites with multi-spectral, hyper-spectral, radar, ultraviolet, and infrared sensors can take "pictures" of almost anything on the planet and in great detail. Different radio waves and light waves can be used to see different things. At one time "spy satellites" had much higher resolution (i.e., could see much greater detail) than other types of sensing satellites. Today there are commercial remote sensing satellites that can see images down to 25–35 cm (about 1 ft) in size. These are not just U. S. governmental systems; there are Russian, Chinese, Indian, European, Japanese, and a growing number of commercial systems like Spot Image (French-based), GeoEye (U. S.-based), etc.

There are some strategic controls, called "shutter control," that that can be imposed on some of the commercial operators that would force them to shut down high resolution active sensing in combat zones during actual conflicts. A sophisticated criminal or techno-terrorist could use the highest resolution data to plan an attack. They could also use data from the new and innovative Sky Box constellation (now owned by Google) that gives rapid updates over a particular spot in order to plot a crime or terrorist attack. This might be to plan a bank robbery when patrol cars are farthest away, or much worse to precisely locate and target the air missile defensive systems that are designed to protect the White House. Although this seems most unlikely, a sophisticated burglar could analyze something like Sky Box data to figure out who is away on summer vacation. In the new world of cyber-crime syndicates and the emergence of the On-line Cyber Criminal Bazaar, as discussed earlier, this use of commercial remote sensing to target homes or offices for burglaries might just become a service to subscribe to through dark channels.

Again, as is the case with all types of applications providing critical services, a higher level of cybersecurity against black hat hacker attack must be provided to make sure that spurious or harmful commands are not sent to these satellites to shut them off or send them into new orbits that would crash into other satellites. Intelsat satellites, which are telecommunications satellites, have not only code-protected commands but have the further protection of having two command stations involved in authenticating the command from an alternative site. One time teen age hackers in New Jersey were able to send a satellite off course, but fortunately the problem was detected and the course corrected before great harm was done [8].

NAVIGATION SATELLITES (KNOWN AS POSITIONING, NAVIGATION AND TIMING SATELLITES). It is navigation satellites such as the American GPS satellite constellation, the Russian Glasnoss, the Chinese Beidou, the Japanese Quasi Zenith, the Indian Regional Navigational Satellite, and the European

Galileo systems that pose the greater risk to Internet security and viability than the previous two types of application satellites. The global synchronization of the Internet currently depends on the GPS system. Cesium atomic clocks on board these satellites (one of the reasons they are so expensive to build) provide time that is exact to a billionth of a second. This system is so exact that the software takes into account Einstein's relativity and the fact that the speed of the satellite must be compensated for even though these satellites move at a tiny fraction of the speed of light [9].

If the GPS satellites were taken out by a particularly violent solar storm (such as the Carrington Event of 1859 or the more recent Montreal Event of 1989) or were attacked by a global jamming operation, the Internet global synchronization could be lost and would be extremely difficult to restore. The potential losses in transportation, banking, retail commerce, etc., could be truly gigantic. This would not of course be a case of techno-terrorists attacking the GPS via jamming. A sufficiently violent solar storm, known as a coronal mass ejection or a X-class solar flare, could also knock out navigation satellites. Each generation of GPS satellite has been designed to be more resilient against jamming or solar storms, but the concern grows as our dependence on navigation satellites grows. It is not just Internet synchronization, but many aircraft takeoffs and landings are also dependent on navigation satellites. The UK, for instance, recently began studies to examine terrestrial backup to Internet synchronization in the event that the GPS system should fail [10] (Fig. 6.3).

Telecommunications Satellites. It is in the area of telecommunications satellites that the issue of cybersecurity truly becomes a truly global problem. This is for several reasons. One reason is that there so many telecommunications satellites out there that could be attacked.

Telecommunications satellites are by far the largest number of satellites currently deployed in orbit. There are quite a few different types of satellites, and quite a few different types of orbits. There are military communications satellites, fixed, mobile, broadcasting, and machine-to-machine data relay satellites. There are low Earth orbit and medium Earth orbit constellations as well as many satellites in geosynchronous orbit (also known as the Clarke orbit, named after the science and science-fiction writer Arthur C. Clarke).

Google had initially proposed deploying high-altitude balloons for Internet communications, in a project known as Loon, and invested in developing a design for such a system. More recently Google has invested $1 billion into Elon Musk's launch company SpaceX. With this move it now it appears that Google might team up with SpaceX to launch a new

**Fig. 6.3** The Global Positioning Satellite (GPS) system constellation, sometimes known as the Bird Cage. (Graphic courtesy of the U. S. Department of Defense.)

constellation (known as the Mega Leo constellation) that might involve up to 4000 satellites, to create just one huge low-Earth system to support low-cost global Internet service. This project in many ways resembles the so-called Teledesic system pioneered by venture capitalist Ed Tuck and initially sponsored by Bill Gates, of Microsoft fame, and other backers from the cellular telecommunications industry, before Tuck's group declared bankruptcy in 2002 [11].

The idea of such a massive system would be provide a huge amount of capacity in low Earth orbit (about 1200 km, or 750 miles) so that there is very little in the way of a transmission delay—unlike satellites in geosynchronous orbit that are out almost a tenth of a way to the Moon and thus require a quarter of a second for a round trip to geosynchronous orbit and back and then a return [12].

The problem of such schemes, however, are numerous. These include the very large and still growing problem of orbital debris. Today over 22,000 sizable objects are already being tracked, and with some 4000 metric tons of debris already in low Earth orbit, there is already radio frequency interference with satellites in close by orbits and with those of other countries and companies in the territory that Musk's constellation would seek to inhabit. The International Telecommunication Union, which is supposed to be in charge of frequent allotments for satellites, the orbital locations, and to coordinate interference, has publicly admitted that it is not equipped to coordinate this proposed new explosion of satellites [13].

SATELLITES OPTIMIZED FOR EQUATORIAL REGIONS AND BROADBAND INTERNET CONNECTIONS. In the last few years there have been a number of new satellite systems designed to provide broadband Internet services to underserved areas of the world. One of the more unique is the system known as O3b. This unusual name stands for the Other Three Billion people that live in the equatorial regions of Earth and which are typically underserved. They are underserved in terms of nourishing food to eat, potable water to drink, clean energy to use, health and education services, and very definitely in terms of broadband communications and networking capability. The O3b system is designed for the equatorial regions and optimized to provide Internet connectivity and links to digital cellular networks. This is a project initially headed by Greg Wyler and is now backed by Google, Liberty Media, and SES Global (the second largest satellite provider after Intelsat). It is Wyler who is the prime architect behind the Google projects to bring new broadband services to the developing world. The system's architecture is also different in that it is a medium Earth orbit constellation of 8 satellites that will expand to 12 satellites. It can be thought of a phase one system that ultimately will lead to the gigantic mega Leo system that Google now appears to be backing [14] (Fig. 6.4).

The good news is that the low latency and Internet-friendly architecture makes this new satellite system conducive to the needs of the equatorial regions of the world. The bad news is that it has no special protections that would prevent cyber criminals to expand their sphere of operations into these regions of the world.

Today the largest provider of digital services designed for the developing economies of the world may well be the Hughes Network Systems (HNS) company that designs very small aperture terminals that work with large geosynchronous satellites to provide rural digital services [15]. Today HNS dominates the supply of interactive VSAT systems around the world, with corporate customers such as Walmart and many developing countries seeking to implement educational, health care, and economic development, among other goals [16].

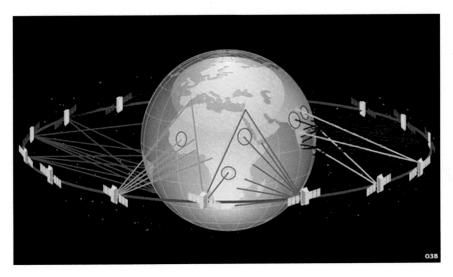

**Fig. 6.4** The O3b satellite network that is optimized for Internet connectivity, lower latency, and low-cost service to the equatorial regions. (Graphic courtesy of O3b.)

About a third of the ATM banking machines in India are deployed in rural India with a satellite link designed by HNS. Although Hughes' systems operate with large and efficient geo satellites, Wyler is trying to lead a campaign with support from Google and others to create new satellite designs optimized for Internet services in rural and remote areas that would offer an alternative to HNS products. The real question is whether these systems have the needed cybersecurity to block criminals from exploiting banking, financial, and credit card services in rural areas, just as they have done for decades in urban and economically developed areas.

## Fixed Satellite Systems and High Throughput Satellites

The most important shift in satellite communications and networking services in recent years has come from what are called "high throughput satellite" systems that are now being deployed such as Intelsat's EPIC, Hughes Network Systems Jupiter, and Via Satellite's impressive nearly 200 gigabit per second satellites. The key to these high performance satellites are multibeam devices that can create small beams allowing intensive reuse of frequencies over and over again in the very broad Ka band spectrum in the 30 and 20 GHz. These satellites can provide very large broadband services to

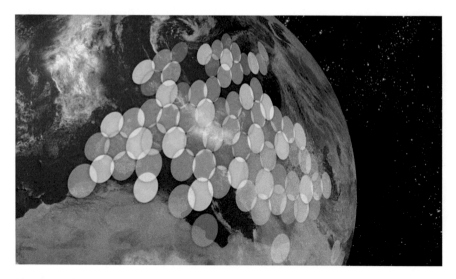

**Fig. 6.5** A beam repetition pattern for a high throughput satellite over Europe. (Graphic courtesy of Eutelsat.)

remote interactive very small aperture terminals (VSATs), not only for business locations but for small offices/home offices (so-called SOHO locations) where fiber is not available.

As we see these innovations in new satellite technologies represented by migration to higher frequencies, more sophisticated antennas able of generating multiple beams, and also advanced coding systems the question that arises as to whether advances in security can also be achieved. The providers at Intelsat General have sought to match Inmarsat General and others in secure end-to-end systems. (Note: The Inmarsat secure service is discussed in the next section.) The coding that allows transmission throughput efficiencies in High Throughput Satellites can indeed also be used to introduce protocols for encryption and security to protect against cyber criminals. On the other hand criminal networks such as within drug cartels can use such technology to carry out illegal operations (Fig. 6.5).

## Mobile Satellite Systems

Perhaps the greatest innovations in satellite services have come in the arena of mobile satellite communications. These are the newest of the big three satellite services and the smallest in terms of total revenues—but the fastest growing revenue-wise. There is amazing new technology in what might be considered "conventional" satellite communications (as represented by systems such as Iridium, Globalstar, Inmarsat, and the Middle Eastern system known as Thuraya) and perhaps even more so in the new hybrid mobile satellites that combine satellite connectivity in rural areas with terrestrial cellular in urban areas.

## Conventional Mobile Satellite Communications

It may be considered odd to refer to some of the most innovative technology ever conceived of as just "conventional" mobile satellite services. But this term is used only to distinguish this type of service from the even newer technology represented by hybrid mobile satellites with ancillary terrestrial components (Fig. 6.6).

The Inmarsat high throughput multi-beam satellites deployed in geosynchronous orbit provide mobile services to ships at sea, aircraft, land mobile, and communications on-the-move services. Its Xpress and Inmarsat IV satellites provide what it markets as a Broadband Global Access Network

**Fig. 6.6** Inmarsat Xpress satellite can provide broadband to vehicles, ships, laptops, and even hand-held I sat mobile receivers. (Graphic courtesy of Inmarsat.)

(BGAN) services. These services can be delivered globally to a variety of ship and aircraft terminals at broadband speeds and to land mobile units at speeds as high as 432 kilobits per second to various terminal sizes. At lower speed it can provide links to even handheld units called Isat phones [17]. Inmarsat is equipped to provide encrypted secure services to support governmental and military mobile connectivity requirements. Its network can provide a wide portfolio of global, seamless, broadband communication solutions and managed network services that are certified to meet governmental security standards. Inmarsat's government BGAN Secure Terrestrial Access (BSTA) solution augments Inmarsat BGAN and other L-band services in order to provide secure full end-to-end network connectivity. This is achieved via Inmarsat's private global Multi-Protocol Label Switched (MPLS) infrastructure in order to offer a secure BGAN terrestrial transport solution.

The Inmarsat state-of-the-art MPLS network provides not only encrypted secure service but high reliability connectivity as well, by using a 24/7 surveillance system and what Inmarsat describes as a proactive management team of security-cleared professionals in the Inmarsat's government Network Operations Security Center (NOSC). These specialized and secured BGAN services are ISO9000-compliant, and all equipment for these governmental and military links is rigorously tested before shipment to government sites [18].

The special feature of the Inmarsat governmental satellite service is that it has a certified secure wireless mobile connectivity that could go from virtually any place on Planet Earth to any other place except the polar regions. There are dedicated military satellite systems such as the U. S. military's Mobile User Operational Satellite (MUOS), but in terms of commercial satellite networks this one of very few options for broadband services. The next generation of Iridium, Globalstar and Thuraya do offer some alternatives, however, and Iridium systems were heavily used by the U. S. military in Iraqi operations as will be explained below (Fig. 6.7).

The narrower band options for secure mobile satellite connections and voice and low data rates are available through Iridium and Globalstar. These feature the low Earth orbit Iridium satellite system and the Globalstar satellite network, which was originally a low Earth orbit satellite system but has transitioned to a combination of different constellation augmentations. Both of these satellite systems support secure military and governmental mobile services as well and provide much lower throughput services for voice and data services to hand held units at the initially quite low data rates of 2.4 and 4.8 kilobits per second, Today's next-generation systems support

**Fig. 6.7** The Isat phone (*left*) and the Inmarsat laptop field office mobile satcom unit on table (*right*). (Images courtesy of Inmarsat.)

higher data rates and a higher quality of service but cannot provide the 432 kilobits per second broader band services available from Inmarsat. The Iridium and Globalstar networks have been used by U. S. military troops in overseas operations in Iraq and other trouble spots around the world. Thuraya geosynchronous satellites have also been used, but on a lesser scale because of security concerns [19].

## Mobile Satellite Systems with ATC Ancillary Terrestrial Component

The great advantage of terrestrial cellular systems is that they concentrate wireless transmission capability into high-density calling areas such as cities and transportation corridors and create small repetitive beams to allow reuse of available spectrum with high efficiency. This terrestrial cellular technology becomes counterproductive and inefficient when the potential users thin out in rural and remote areas. This is where mobile satellite systems with larger beams and much greater geographic coverage become effective and efficient.

The Terrestar satellite system as depicted below is paired with terrestrial cellular technology to provide broadband mobile services for the entire U. S. geographical area and significant areas of Canada. This is the type of satellite that the U. S. Federal Communications Commission (FCC) has licensed, to provide the mobile satellite service with an ancillary terrestrial component, which means it is a hybrid space and terrestrial mobile service for the entire United States on an integrated basis. A similar type of service is also foreseen for Europe, even though the service there is known there as mobile satellite service with complementary ground component.

There was another system designed and deployed for the United States known as Light Squared that launched the SkyTerra satellite for this type of hybrid mobile service. Although this service was initially licensed by the FCC, it turned out that the frequencies assigned for the terrestrial component were too close to the GPS frequencies, and as such would cause too much interference to the navigational and timing service. This led to the service being put into limbo and resulted in the bankruptcy of this particular venture [20].

In terms of cybersecurity issues the MSS-ATC services are very much the same as terrestrial broadband cellular, and the same precautions that were discussed in the discussion of the mobile cellular service apply. The ability to listen into and obtain data such as credit card information or electronic banking records from an anonymous and protected site are even greater with the combined satellite and terrestrial networks (Fig. 6.8).

**Fig. 6.8** The Terrestar satellite with a gigantic deployable multi-beam satellite. (Graphic courtesy of the Harris Corporation.)

# Conclusions

The answer to providing SCADA remote control security is really not all that difficult. We need to make sure that the access codes to these systems are reasonably intricate and secure, periodically updated, and there is an independent security audit to determine that all of these measures are being carried out. It is highly recommended that a technically sophisticated agency be charged with enforcing cybersecurity on SCADA networks rather than an operating agency such as the water department, electrical or gas utility unit, or traffic or transportation department. These functional elements are typically focused on during the prime job and not necessarily the best overseeing unit for cybersecurity.

The question of satellite security is a much more difficult and hard to address problem. There are just too many different types of satellites, in different types of orbits, using different frequency bands, with too many coding schemes, and too many user interface protocols for there to be any systematic security controls on a national, regional, or global scale. Some 200 countries and territories now use satellites for weather forecasting, remote sensing, navigation and timing, and especially telecommunications and IT networking. Over 50 countries have deployed and are independently operating application satellites, and there are thousands of companies either designing, operating, building, or providing content and control systems for information that is transmitted via satellites.

And if the problem today is difficult and complex, just think of the new projects involving hundreds or even thousands of satellites in a single constellation that could provide Internet access from anyplace at any time. On top of presenting cybersecurity issues, it could make the orbital space debris problem that has gotten steadily worse exponentially worse still. Decades ago Dr. Donald Kessler of NASA predicted the possibility that orbital debris could grow to the point where it could become self-sustaining and spiral out of control. This prediction, now known as the Kessler Syndrome, seems likely to come true unless new controls are put in place beyond the voluntary U. N. guidelines that have only served to modestly reduce the rate of orbital debris build-up.

Certainly, there are encryption systems that can provide security against eavesdropping and also to protect against critical space infrastructure cyberattacks. The George C. Marshall and the Tech America Space Enterprise Council recently expressed its concern of unprotected vital space resources in the following manner: "The evolution and increased complexity of information technology capabilities, which are a key component of space systems architectures, have made these systems more vulnerable to cyber-attacks. The growing concern over cyber threats has made us focus more intently on mission resilience" [21].

There is also a need for improved IP security standards (IP Sec) that allows them to operate over satellites and to compensate for satellite delay that creates difficulties for efficient satellite transmissions—especially for geosynchronous satellites. Nevertheless there are a variety of security issues, orbital debris concerns, frequency and interference coordination, and space traffic management issues that will need to be addressed, and sooner rather than later.

As we enter a new world with more commercial space launching capabilities, more large-scale satellite constellations, more demand for communications and networking satellites to reach underserved portions of the world, the problem of cybersecurity on space systems will also increase. One of the newer concerns is the current reliance of the global Internet on GPS satellites for system synchronization and what we would do if an extreme solar storm or terrorist attack were to disable or greatly reduce the effectiveness of our space-based positioning, navigation and timing systems.

# References

1. Chris Frates and Curt Devine, CNN, December 12, 2014 http://www.cnn.com/2014/12/19/politics/government-hacks-and-security-breaches-skyrocket/index.htm.
2. U. S. General Accounting Office Report to Congressional Committees, "Library of Congress: Strong Leadership Needed to Address Serious Information Technology Management Weaknesses," March 2015. http://www.gao.gov/assets/670/669367.pdf.
3. Joseph N. Pelton, Indu Singh and Elena Sitnikova, "Cyber Threats, Extreme Solar Events and EMPs," *Inside Homeland Security*, March 2015.
4. Indu Singh and Joseph N. Pelton, *The Safe City: Living Free in a Dangerous World*, (2013). Emerald Planet, Washington, D. C., pp. 199-200.
5. Anggoro K. Widiawan, and Rahim Tafazolli, "High Altitude Platform Station (HAPS): A Review of New Infrastructure Development for Future Wireless Communications." Wireless Personal Communications August 2007, Volume 42, Issue 3 2006. pp 387-404.
6. Joseph N. Pelton, Scott Madry and Sergio Comacho-Lara, *Handbook of Satellite Applications*, (2013) Springer Press, New York, Chapter 1.
7. Joseph N. Pelton, "Meteorological Satellites," in Joseph N. Pelton's, Scott Madry's and Sergio Comacho-Lara's *Handbook of Satellite Applications*, (2013). Springer Press, New York.

8. Siamak Khorram, Frank H. Koch, Cynthia F. van der Wiele, and Stacy Nelson, *Remote Sensing* (2012). Springer Press, New York.

9. Scott Madry, *Global Navigation Satellite Systems and Their Applications* (2015). Springer Press, New York.

10. *op cit.* Joseph N. Pelton, Indu Singh and Elena Sitnikova, "Cyber Threats, Extreme Solar Events and EMPs," *Inside Homeland Security*, March 2015.

11. Joel Hruska, "Elon Musk unveils new plan for global satellite Internet, while Google invests a billion in SpaceX," ExtremeTech, January 20, 2015. www.extremetech. com/extreme/197711-elon-musk-unveils-new-plan-to-circle-to-earth-in-satellites-for-fast-low-latency-internet.

12. Elizabeth Palermo, "Google Invests Billions on Satellites to Expand Internet Access," Live Science, June 4, 2014. http://www.livescience.com/46109-google-satellites-expand-internet-access.html.

13. Joseph N. Pelton, Orbital Debris and Other Threats from Space, (2013) Springer Press, New York.

14. O3b Networks. www.o3bnetworks.com/welcome-to-o3b.

15. Hughes: Broadband Satellite Services, Managed Networks www.hughes.com/.

16. The Comsys VSAT Report, 13th edition, "Hughes Market Summary and Company Profile," London, UK.

17. Inmarsat Global Xpress. The first high-speed broadband network to span the world http://www.inmarsat.com/service-group/global-xpress/.

18. *Ibid.*

19. Ramesh Gupta and Dan Swearingen, "Mobile Satellite Communications Markets: Dynamics and Trends" in Joseph N. Pelton's, Scott Madry's, and Sergio Comacho-Lara's (eds.) *Handbook of Satellite Applications* (2013). Springer Press, New York, Chapter 8, pp. 163-186.

20. *Ibid.*

21. "U. S. Cyber and Space Security Challenges and Opportunities," George C. Marshall Institute and Tech American Space Enterprise Council Symposia, April 10, 2015.

# 7

# Who Will Control the Future, Black Hat Hackers or the Hacked?

## The Viral Cyber World

We now live in what might be called a "viral cyber world." Disruptive changes are constantly occurring in this viral cyber world. Such changes do not track a simple linear upward pathway nor even expand at what might be called an exponential rate. In our rapidly changing viral cyber world, change is nearly instantaneous and explosive.

People today often have hundreds—perhaps even thousands—of friends or followers on Instagram, Facebook, LinkedIn, or Twitter. They can pass along something they see on the Internet in an instant. "Viral explosions" on the Internet can zoom across the world in seconds. This digital explosion can be a cute video, a political faux pas, or something much worse. It can include a new devastating virus, malware, or totally untrue information that could cripple the career of a politician, a leading executive, or a Hollywood star. It can include false or misleading information that can send a stock price into a nose dive. Erroneous information about a private jet crash that killed a key executive of a Fortune 500 company is more than a prank. In the world of 24-hour-a-day cable television and runaway bloggers, such a bogus report could create millions of dollars of losses and criminal charges against perpetrators.

Such a viral explosion within a small group such as a school community can constitute a major source of bullying. Within an office it can be a shaming or shocking experience that depicts a lurid image of someone that has been shamelessly photo-shopped. A cut and a paste and a clip and voilà, you can be shown online as doing almost anything. You can appear to be bashing someone's head in, or perhaps reeling as if you were stone drunk. Maybe you are shown to be kissing someone that you shouldn't, or even appear to

© Springer International Publishing Switzerland 2015
J.N. Pelton, I.B. Singh, *Digital Defense*, DOI 10.1007/978-3-319-19953-5_7

be completely and shamelessly nude or even having sex. Later you might be able to set the record straight, but a great deal of damage might never be undone. The future, unfortunately could eventually even be worse.

The cyber world has the power to accomplish much that is good—but much that is bad as well. It can spread knowledge and wisdom or racist or extremist political views. At its worse it can be an instrument of terrorist and criminal acts. One of its defining characteristics is incredible speed. Electronic systems can encircle the globe in an instant and in the world of social media, politics, finance, and military defense such speed can be dangerous—especially if the information is fraudulent, misleading, slanderous, criminal, or intended to initiate a terrorist attack.

In the world of physics there is velocity (distance/time). Then there is an increasing rate of velocity that is acceleration (distance/time$^2$). Then there is an explosive force that physicists call a "jerk." This is where the rate of acceleration is itself increasing and is expressed mathematically as a third or fourth order exponential. In nature when we experience something that is mathematically expressed as a third or fourth order exponential it is usually a very bad thing. Something that can be characterized as a 'jerk' is a phenomenal explosion such as the Big Bang or a nuclear explosion or at the very least what happens when your head snaps back as someone slams on an accelerator. Before the Internet and Facebook we had no means to experience explosive social media effects that were the equivalent of 'jerk,' but now we can. People can start an avalanche of information (whether true, false, infectious, or dangerous) that can and does spread around the world—perhaps anonymously and often perniciously—and in just a few minutes. This is today's reality. This capability, in short, can be powerful, instantaneous, anonymous, dangerous, global in impact, and if targeted toward critical infrastructure such as trains, planes, power plants or water supplies potentially devastating in impact.

## How Do We Protect Ourselves from Cyber-Attacks Going Forward?

Our world has changed and our ability to protect ourselves against harm has now taken on a new dimension. Today we need not only physical security to stop people with weapons or bombs from attacking us but cybersecurity as well. We need cybersecurity to protect ourselves, our family, our communities, and our nations. This protection is needed against digital errors and mistakes that can happen by accident or negligence or against natural disasters that could destroy our vital digital and electronic infrastructure on which we depend. Other forms of protection are needed against criminal activity, against cyber-terrorists, and against cyber-attacks by other nations. The truth is that we live in a world today that is not only different from the time of cavemen but radically different from even the time of eighteenth and nineteenth century industrialization. In our world, where 80 % of all people in developed economies depend on jobs in the service sector and 3 % of all people are engaged in farming and mining, we are quite vulnerable if the lights go out, the power fails, or our satellites and networks are put out of commission.

In the last 50 years we have increasingly created a world that is many ways more desirable. It is prosperous, more comfortable, less exposed to virulent diseases, better housed and fed, and so on. The modern world is, however, much more vulnerable, because we depend so much on critical infrastructure and information networks that run our machines, store our knowledge, and allow us to leverage a highly mechanized and automated world. If our automated world breaks down we become quite vulnerable. This vulnerability increases as we add more and more automated controls, yet we are not trained nor equipped to function without our intellectual infrastructure instantly at hand to command.

A natural catastrophe such as an asteroid strike, a coronal mass ejection from the Sun, or some other huge event could create massive destruction. These events could wipe out our power, transportation and food and water distribution systems, and our communications and information technology systems as well. It could result in the deaths of over a billion people. Lloyds of London has conducted a study that concluded that if a massive solar storm hit Earth and wiped out our electrical power systems, our satellites, and shut down our transportation and communications systems that the estimated economic harm would be $2.6 trillion (U. S. dollars) This is the economic equivalent to 25 Hurricane Katrinas or 40 tropical storms like Sandy, and the deaths would be even proportionately greater. As the speed of our communications and information networks speeds up so does our potential vulnerability [1].

You would be right in thinking an asteroid strike or solar event would surely not kill a billion people immediately. No, the deaths would be lingering and sustained over the months that followed. If our electrical power grids and transportation networks shut down for any length of time, however, hundreds of millions of people could die for the lack of food, water, heating, or even starve because they would no longer have a way to make a living. We are no longer organized for subsistence living. People simply do not realize the many ways we are today dependent on our power and electronic systems and software [2].

Today people are, for instance, more and more dependent on credit cards. And that dependency will only increase in the age of cyber-currencies like "Bitcoin" or the newly very popular "Apple Pay" or "Google Wallet."

National governments have yet to make a clear-cut and definitive assessment of what would happen if all our electronic money systems, all our bank accounts, and all of our stock market holdings were wiped out by a natural disaster, or far more likely, by a terrorist-sponsored attack. Further there is no guarantee that criminal or cyber-terrorist attacks by increasingly sophisticated 'crackers' will not be able to continue to exploit weaknesses in cybersecurity to put your assets at risk.

In the world of counter intelligence, there is the world of counter-counter intelligence, and even counter, counter, counter intelligence. In the world of cybersecurity there is most definitely the world of counter cybersecurity. Cyber criminals and cyber-terrorists are increasingly able to counter efforts to control their attacks.

The security system used on the World Wide Web for protected financial transactions is today largely based on so-called "public-keyed infrastructure." The idea here is that a public key (a very long and random list of numbers, letters, and symbols) would be combined with a private key (again an individually created long list of numbers, letters, and symbols). The issuance of a public key would come in the form of a "certificate" from a "protected high level organization" that is thought to be reliable. The trouble is that more and more organizations such as Microsoft, who have become involved in issuing certificates, are now subject to attack. Some overseas banks use a public key for a long time, and their security for private keys can become suspect. In short, the problems with banking security and public keyed infrastructure (PKI) have mounted over time.

One of the problems with the PKI system involves the so-called X.509 standards related to the issuance of digital certificates known as CAs:

The most sophisticated attackers have recognized the importance of digital certs and are targeting the root of this trust, the CAs. Malware such as Stuxnet and Flame are being signed, to bypass system security checks. This means that the explosion of the use of "certs" within the organization has introduced a vulnerability, one that needs to be managed, starting with discovery of that vulnerability. As we move to a more advanced form of encryption, there are cyber criminals seeking to break down the encryption or otherwise defeat the cybersecurity that is used by banks, retailers or those involved in e-commerce [3].

The recent disclosure by the U. S.-based Anthem Health Care System that some 80 million customer files were compromised has led to calls to create a new layer in the Internet Protocol to include a "security level" that would verify that a message is coming from a secure and bona fide site in order to protect Internet users. The trouble is that, as just discussed above, there are ways to get "bogus certificates," and thus messages from such sites would still be able to come through.

The problem of false certificates is just one example of the ways we are vulnerable to a cyber-attack in today's complex digital world. It was recently revealed by Kaspersky Labs, the Russian cybersecurity firm, that black hat hackers have been able to infiltrate the security systems of 100 banks using simple phishing schemes and similar methods to find out these bank's operating systems. With this information in hand they were able to dispense cash through ATM machines over a matter of months to accomplices. From 2013 into 2015 this ring of cyber thieves was ultimately able to realize a billion (U.S.) dollars in e-robberies, even though they limited their ill-gotten gains to $10 million at any one particular bank. In short, as one security gap is closed another one will surely open. Consumers like you and me will ultimately pay for these crimes as banks and financial institutions pass on their losses to their clients [4].

The fundamental problem is that our modern world is composed of a series of what might be called a 'one way gate' that innovations open one by one but at an ever increasing speed. As we invent transistors, monolithic computer chips, satellites, vast global electronic networks, lasers, and yes, even spandex and the pill, we cannot simply just 'uninvent' them. Advanced technology begets even more advanced technology. We are at our very essence technological beings—a race of inventors. We have invented the Internet, smart phones, Wi-Fi wireless LANs, and a world that runs on electronic communications, but we have yet to find a way to disable or prevent criminal or terrorist use of these networks. The very efficiency of these electronic networks now represents a vulnerability when it comes to cybersecurity.

# Coping with Current and Future Cyber Threats

The threat to one's cybersecurity is today constantly at issue. There is the issue of computer fraud that can harm you in a wide variety of ways. Criminal hackers can steal your computer card or electronic checking information or your account number and private code at your stock broker. Cyber criminals can thus endanger the security of your bank account, your stock portfolio, or create problems for you on social media. You can be portrayed as morally loose or as committing criminal acts. Your Twitter account can be hijacked to send out embarrassing, obscene or even compromising messages that would appear to be confessing to crimes or horrible deeds. Even if you do not have a major attack simple cyber-tricks can be an irritation. Such tricks might end up with you signed up for a magazine subscription or purchasing products you did not want to buy.

Identity theft and related misuse continues to be on the upswing despite computer security systems that you can purchase to protect your computer security. It is possible to purchase insurance protection against a number of cyber-risks that include data breaches, identity theft, financial losses, network damage, and cyber extortion. Despite the reality of the threat only a percentage of Internet users actually have such insurance protection. Most individuals tend to have computer virus protection, and some have both firewall and virus protection that includes some insurance protection, but millions are still at risk, even if one does have computer system protection at work and reasonable protection at home. The risks do not stop there.

## Nearer Term Threats: Computer Fraud, Wi-Fi, Wireless LANs, and the Cloud

Beyond computer fraud, malware, spyware, and types of cyber-attacks, there are other things to watch out for in your everyday life as you go about your everyday routine. If you use a publicly available Wi-Fi network any communications you conduct are vulnerable to interception. If you have a wireless LAN in your home it is subject to interception. If you access the Internet via a broadband satellite data link you also need to take precautions with regard to the use of virtual private networks. In satellite transmission systems that have not been optimized for Internet service the IP security header that provides the Internet address for data packets can be stripped off in the transmission process. Fortunately most satellite networks have now been optimized for efficient data transmission, and IP security now works well via satellite connections.

Even if you have a protective code on your wireless network at home, if the ID happens to be your birthdate, your Social Security number or your name, your middle name or the name of your spouse, it might not take too long for someone with a scanner to break your secret code. The simple truth is that if you can avoid using a wireless router in your home and provide direct cable networks, this is the safest way to go.

# Wireless Local Area Networks (Lans) and Satellite Networking

It stands to reason that electronic radio wave transmissions that travel through the air are by far the easiest to intercept. This is not to say that transmission via copper wire, coaxial cable, or fiber optic networks are immune to interception. But trying to intercept data signals from 'hard wire' systems requires super sophisticated James Bond-type technology such as using sophisticated limpet systems that government spy agencies might employ. Thus any time you rely on satellite or wireless LANS or digital cell phone service you do not necessarily have protection. Even if you have passwords on your Wi-Fi, wireless LAN, or other wireless device you could still be vulnerable. Only if you have a digitally encrypted 64-bit or above system can you feel completely protected.

Wireless LANs and Wi-Fi systems are a problem in many ways. With a scanner and recorder wireless traffic can be intercepted and taped. Without you knowing it an eavesdropper can capture passwords, learn how to access your financial accounts and much more. It is not exceptionally difficult to get access to your wireless LAN or a Wi-Fi system to transmit spam, or use your network to perhaps launch an attack against others. Black hat hackers can capture and modify traffic to masquerade as you. In this guise they can seek money or engage in scams of various types. This type of identity theft can have serious legal consequences. Even a low-tech attacker can disrupt your business by what is called a denial of service by sending of flood of packets into a particular website.

It is important within an office to frequently carry out WLAN site surveys via a WLAN discovery tool such as Win32 NetStumbler or Meraki's Java Cloud Stumbler. This process will turn up any unauthorized workstations. It is important to create an inventory of laptops and PDAs with wireless adapters to document users. This process will be used to implement WLAN access controls. It is particularly important to establish an up-to-date list when WLAN adapters are lost or stolen. Also it necessary to sweep the office in the pertinent frequencies to locate unauthorized access points (APs) that might be connected to your offices via wireless LANs. Possible tools might include Wi-Fi Analyzer, Heatmapper, Kismet, Wireshark, Nmap, WiFiDEnum, Aircrack-ng, MDK3, and Karmetasploit. This process can identify bogus network names (known as Service Set Identifiers, or SSIDs) that are surreptitiously able to log on to the wireless network.

Finally vendor default codes are another security concern. It is important when new wireless equipment is obtained to change the default security code. This is a problem for telephone exchange equipment, wireless LANS, and any electronic equipment with a default security code [5].

## Security Flaws in The Cloud

The providers of computer Cloud services provide access to software as well as platforms for data processing and storage. These are known as providers of software as a service (SaaS) and providers of a platform as a service (PaaS). Obtaining software through The Cloud or obtaining processing power and data storage can lead to cybersecurity breaches.

IBM, Cisco, SAP, EMC, and several other leading technology companies announced in late March that they had created an 'Open Cloud Manifesto' calling for more consistent security and monitoring of Cloud services. But the fact that neither Amazon.com, Google.com, nor Salesforce.com agreed to take part is indicative of the problem. It suggests that any universal industry consensus is far from certain. Microsoft also abstained, charging that IBM was forcing its agenda: "Standards by definition are restrictive. Consequently, people are questioning whether cloud computing can benefit from standardization at this stage of market development," says Trifković. "There is a slight reluctance on the part of cloud providers to create standards before the market landscape is fully formed." [6]

## Best Practices for Companies in the Cloud

- Use agentless exception monitoring systems to identify system errors. Such system errors may in fact indicate or involve security breaches.
- All updates should be carefully monitored. It is important that updates do not enable staff to achieve unwarranted access. The massive data breach achieved by Edward Snowden is just one illustration of unwarranted data breaches. It is prudent to separate functions so that access systems are segregated rather than accessible by all system analysts.
- It is important that as data is placed on a Cloud that there is a clear certification as to where it is stored and in what country and locale. This includes knowledge as to the applicable data protection laws in the relevant jurisdictions.
- Seek a periodic and thorough independent security audit of the hosting site for The Cloud.
- One of the objectives of the audit would be to find out which third parties the Cloud company has business relations with and whether it is able to access your data and what certification exists that your data is firmly segregated.

- Password protection is key. There should be precise procedures and protocols as to how passwords are created, protected, and changed, and who has authority to do this and processes for periodic change of key passwords.
- The audit should also test and confirm availability guarantees and clearly identify penalties for noncompliance.
- It is important that a Cloud provider agrees in writing as to being able to accommodate the specific security policies that you have put in place. Paying a premium to ensure that all of these security practices can be observed, authenticated and audited [7].

# Cyber Protection of Vital Infrastructure Under U.S. Government and Corporate Management

President Obama issued Executive Order 13636 (EO), "Improving Critical Infrastructure Cybersecurity," on February 12, 2013. This executive order was issued after attempts to enact mandatory legislation in the U. S. Congress were not successful. Exactly a year later a new framework was published in the form of a 40-page White Paper that had been developed by the U. S. U.S. National Institute of Standards and Technology (NIST) in response to this executive order. The White Paper, entitled "Framework for Improving Critical Infrastructure Cybersecurity," was developed in cooperation with industry experts. It is in essence a cooperative Cyber Security Framework (CSF) that allows a common process and set of communications protocols that can work for small, medium, and large organizations to develop cybersecurity processes to identify threats, protect vital databases, detect intrusions, respond to such intrusions in an effective manner, and recover from any lingering impacts of such intrusions while also improving security [8] (Fig. 7.1).

It is important to take this general framework and break it down into key steps or detailed categories such as has been undertaken by Los Alamos Technical Associates to translate the framework into an actual security program that can be systematically implemented [9].

Lisa Monaco, who is President Obama's counterterrorism advisor, has described document as providing "a common language to discuss cybersecurity" with private and federal agencies. As reported by Cameron Camp, there has been talk of cyber risk insurance providers using the framework as a benchmark when determining rates. If a company could show it is conforming with the framework it could potentially buy cyber insurance for less, just as insurance for a dwelling costs less if the building is built to code. However, Cobb says Sedgewick took great pains to point out that the

| FUNCTION | CATEGORIES |
|---|---|
| IDENTIFY | Asset Management |
| | Business Environment |
| | Governance |
| | Risk Assessment |
| | Risk Management |
| PROTECT | Access Control |
| | Awareness and Training |
| | Data Security |
| | Information Protection Processes and Procedures |
| | Protective Technology |
| DETECT | Anomalies and Events |
| | Security Continuous Monitoring |
| | Detection Processes |
| RESPOND | Communications |
| | Analysis |
| | Mitigation |
| | Improvements |
| RECOVER | Recovery Planning |
| | Improvements |
| | Communications |

**Fig. 7.1** The NIST cybersecurity framework. (Graphic courtesy of Los Alamos Technical Associates-LATA.)

framework was not intended to be the basis for mandatory standards, a view underlined by the president's uses of the term "voluntary framework."

In the wake of the Sony and Anthem data intrusions, President Obama a year later in February 2015 moved even further to create a new overall unit that will provide a single point to alert businesses and all federal agencies to imminent cyber-attacks or help coordinate actions if and when new attacks are detected. Part of the mandate of the new unit will be to speed up the sharing of data between and among the intelligence agencies, the rest of the federal agencies, state and local governments, and private enterprise.

This new agency that was announced by President Obama on February 10, 2015, will be charged with responsibility for providing a coordinated analysis of cyber threats, modeled on similar U. S. government efforts to fight terrorism. Critics of this new unit, such as Tom Kellerman, chief cybersecurity officer at Trend Micro, Inc., have noted that there is a unit within the Department of Homeland Security that is currently charged with such a responsibility and that this might be adding a new layer of bureaucracy but without enhancing cybersecurity threats.

Lisa Monaco, however, has indicated that the new unit will have a broader mandate and global perspective as well as more effective links to intelligence agencies in terms of not only rapidly pooling and disseminating data on cyber-breaches but also in creating new tools that seek to anticipate attacks as well.

Specifically Monaco said: "Currently, no single government entity is responsible for producing coordinated cyber threat assessments" and that this action was urgently needed because the number and seriousness of such attacks "are ballooning in size" [10]. The new agency, the Cyber Threat Intelligence Integration Center, "is intended to fill these gaps," she said.

It is hoped that this new unit will be well integrated into the new U. S.-NIST-Developed Framework Core System by Cyber-Security and provide the essential first function to "identify" the threat. Figure 7.2 below indicates how the overall process of risk identification, risk management and risk protection implementation would likely work [11].

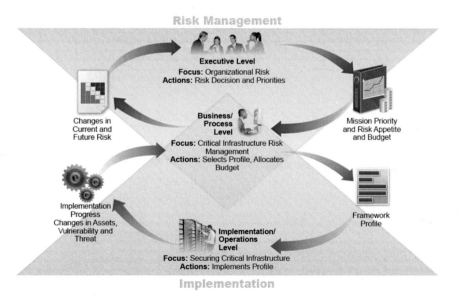

**Fig. 7.2** Layers of protection and cybersecurity action in the U. S. framework. (Graphics from NIST report on cybersecurity framework.)

# The Smart Grid

The electrical grids around the world are antiquated and subject to failure. A failure of one part of the grid due to lightning strikes or a major solar event can trip failures in other regions one after another in a cascade effect. Alarms are supposed to sound and automatic switches are supposed to stop such widespread cascading power failures, but in 2012 in India the power in the north failed so that 40 % of the country's population was without power. In the United States and in other countries massive new investments are being undertaken to create smart grids that can conserve power, make them greener, and less likely to fail in this cascading manner when there is an overload in one particular area. There are many advantages of installing smart grids and district power systems that makes power consumption more efficient and forestalls the need to create more energy generation.

There is one disadvantage of smart grids. This is that while grids are more secure and more protected against failures from natural causes, black hat hackers could potentially find ways to create cyber-breaches in the power networks and manipulate them for criminal or even terrorist purposes. This means that the $3 billion that the United States is spending to create and install the "smart grid" across the country includes concerted efforts to enhance the security against cyber-attack. Thus the new cybersecurity framework that has been developed by NIST pays particular attention to using this process to protect the smart grid. Efforts include the use of more elaborate codes that are changed more frequently and dividing responsibilities so that one person is capable of sabotaging entire systems.

Also there is more and more independent auditing of security measures, and particular attention is being paid to the security of SCADA systems that are key to today's large automated networks for power, traffic lights, water and sewage treatment systems, etc. As we move to the "smart grid" and the "smart city" through automation and artificially intelligent control systems, we also are seeking to move toward the "safe city." In this "safe city" there will be more encryption and more security applied to access systems so that unauthorized hackers cannot use the capabilities of smart systems against us [12].

# Vulnerable Apps on Smart Phones and Desktop Computers

The breathtaking spread of applications for computers and smart phones is truly astonishing. Every retailer, every newspaper and magazine, every college and university, every bank and credit card company has developed a convenient app to let you shop, compare, access, apply for admission, or carry out a financial transaction. Most apps are easy to install and involve little or no cost to put on your mobile or desktop computer, laptop or smart phone. But depending on which platform the application was built on you may be inviting a cybercriminal to steal your passwords, access your electronic banking, or even carry out identity theft.

In 2008 the worries were more about applications that one loaded on their desktop, and Zero Day listed the top 12 "apps" that had required the most patches to prevent some form of cyber fraud. The list way back then started with Mozilla Firefox as the most "patched" app followed by Adobe Flash and Adobe Acrobat, EMC VMwarePlayer/Workstation, Sun Java JK and JRE, Apple QuickTime, Safari and iTunes, Symantec Norton products (all flavors), Trend Micro OfficeScan, Citrix Products, Aurigma Image Uploader/Lycos File Uploads, Skype, Yahoo Assistant, and Microsoft Windows Live (MSN) Messenger. The chances are that virtually every user in the United States if not the world was vulnerable through one of these apps [13].

Of course in today's world, with billions and billions of cell phones, including several billion smart phones loaded with scads of applications, the vulnerabilities just keep escalating. The dangers of using the apps on your I-Phone or Android for other than games and communications (text or e-mail) are, in a word, SIGNIFICANT. One strategy is to upgrade your password with the latest software. One can go to either the I-Phone site or a Samsung Android site and upload a system that uses your face to open your cellphone (Fig. 7.3).

One's password might be discovered, but your face is a one of a kind. Apple first patented a system of this kind on September 20, 2012, but Android and others quickly followed suit with their own inventions. The initial software in 2012 was not very sophisticated, but today's systems are quite good, and this does not need to be constantly shifted around and remembered [14, 15].

If you are using your smart phone for online banking or things like Apple Pay, our advice is threefold: (1) Don't do it. (2) Change your passwords frequently or shift to an access unlocking system based on your face. Even so don't expose your wireless transactions to accounts with large amounts of money. In short, keep your larger account segregated from smart phone transactions. (3) Talk to a trusted advisor at a bank or financial institution about how to protect your financial transactions. This is because the dangers of some sort of application loaded on your computer will be used against you as problems with cybersecurity applications continue to increase.

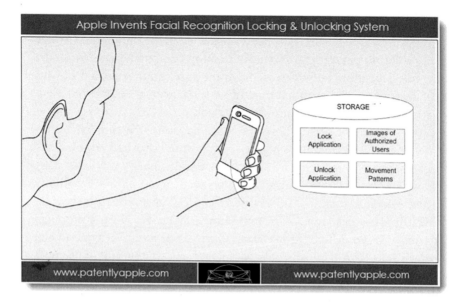

**Fig. 7.3** Facial recognition to unlock your cell phone by Apple. (Graphic courtesy of Apple.)

# The Internet of Things

Another future trend beyond automation, SCADA control systems, artificial intelligence in large networks, and encryption in smart grids and networks. is what is now called the Internet of Things. In the first days of computers were the large main frame computers that only governments and businesses could afford. Thomas Watson, the Chairman and CEO of IBM, was alleged to have said in 1958: "I think there is a world market for about five computers." The invention of transistors, integrated circuits, monolithic devices, and high capacity digital processors on a chip has, of course, changed the world and has allowed the widespread growth of desktop computers, laptops, and even smart phones that have far more processing power than the earliest computer mainframes.

Today almost everything that is sold that might include some form of intelligence to control functions contains computer chips that can process information and receive and transmit information. Automobiles, trucks, buses, trains, airplanes, refrigerators, washing machines, dishwashers, computers, laptops, cameras, scanners, printers, smart phones, smart wrist watches, television sets, radios and disc-players, game consoles, sewing machines, thermostats, burglar alarms, and perhaps even ovens, bathtubs, and skill saws have or will have their own digital processors and ability to communicate with their users and the Internet. Manufacturers can send software updates to various appliances so that television sets can receive higher definition signals. Showers may receive updates so that they can cover your body better and stronger but with 30 % less power. Cars can tell police and fire fighters and rescue workers that you have had an accident and need help. This is the good news that comes with the world of appliances and vehicles getting smarter and more able to communicate.

The bad news is that so-called "spammers" could take over your refrigerator and start sending out thousands of unwanted spam messages. It could mean that an assassin might communicate with your smart fuel injector system so that your car goes out of control on a deadly mountain road. The key question that now arises is, should manufacturers be able to communicate with automobiles and appliances with "going through the owner first" and are owners going to be able to encode their own smart appliances so that in this world known as the Internet of Things you will be in control rather than manufacturers, the government, or, worst of all, black hat hackers or even terrorists? These are issues that legislators are going to have to take on sooner rather than later [16].

## Hololens and Cybersecurity

There is a new CEO of Microsoft, Satya Nadella, who is trying to reawaken the slumbering giant that is now Microsoft. He aspires to create a future within just a few years that today seems very much like science fiction. Nadella's vision is not only to spread Microsoft Windows across all platforms, including iOS and Android, but to create a whole new interface for all modern IT users that is intuitive and incredibly easy to use.

The idea is to facilitate what amounts to a virtual reality link between users and whatever they wish to experience—whether it is to learn a new skill, experience an artistic performance as if you were a participant, play an interactive game, store vital information, or process data in a way that is compellingly vivid, or see the equivalent of a holographic movie. One will only have to put on a holographic google to interface with a wide range of experiences that is useful for learning, for stimulating thought, or for having an interactive gaming or entertainment encounter.

As realistic holograms, your digital content will be as real as physical objects in the room. For the first time, depicted holograms could become practical or whimsical tools of daily life in schools, in shopping malls, or a new form of X-rated entertainment.

With the new complex software being shaped by Microsoft developers one can use the "hololens" interface capability to fine-tune a design, learn how to fix a light switch, or cook a soufflé. You can interact with hologram images to learn something new. You can use this technology to share your ideas or to show and tell your business proposition from multiple perspectives. Microsoft holoLens technology of the near future could help you make decisions more confidently, work more effectively, and bring your ideas to life right before your eyes by blending hologram with reality.

It is always hard to forecast the future, but it seems highly likely that the world of interaction with computers and mobile devices will move beyond having a screen to touch, a mouse to click or a Siri to talk to. With hololens-type technology it would in theory be possible to use gestures or even eye moments to create, shape, or alter the size of holograms and transfer, edit, or transform information. In the world of the future it will be possible to use your mouth or eyes or hands and arms to navigate, explore, and communicate with your apps. The next step in the cyber world's evolving nature is to convey content and information in the most natural way possible, and if Microsoft's latest development proves successful, then this will be via interactive holograms.

As is always the case with new technology there are downsides, and in this case it will most likely to be in the area of cybersecurity. The challenges of the mega data interchanges involved with holographic interchanges and the potential loss of privacy and cyber-attack are probably north of ENORMOUS. There is great potential for black hat hackers to penetrate the super-broad interfaces involved in hologram imaging. This potential breach of security would be difficult to block. The ability will be there even more than before to plant listening device bugs, viruses, Trojans, data bombs, or steal vital corporate or even secure governmental information [17].

## Conclusions

The future will be dominated by an even increasing onrush of modern technology. As we noted earlier, the invention of new IT technology—wireless LANS, satellite and wireless virtual private networks, Cloud-based processing and information storage, e-money, the Internet of Things, and holoLens interface experiences—all are exciting, interesting, and dangerous. For the most part this new technology is a one-way gate to the future. Scientists, engineers, inventors, and marketers are constantly providing us with new capabilities and opportunities in a never-ending stream. This will be the case unless we enter a cataclysmic new Dark Age.

New technology such as an Internet of Things and communications via holographic images and memory systems cannot just be "un-invented" when we realize that these new technologies bring with them the Trojan horse of cybersecurity dangers. The problem of cyber-attacks, invasion of privacy, stolen data, and identity theft will become ever more difficult to block, because the torrent of information exchange will only become larger and faster and more internationally diverse. Cyber criminals and cyber-terrorists will find that new and faster broadband information technology will bring yet new opportunity to attack financial institutions, your individual banking accounts, your Social Security account, and much more. As we move to the age of quantum computers that process information at petabytes per second speeds and communications with ever appealing and personalized holographic images, the victims will likely be your privacy and your cybersecurity.

## Post-conclusions: Cyber Opportunities vs. Cyber Threats to Our Lives?

The new cyber world that unfolds ever more rapidly each day will bring new opportunity for education, training, business, global economic exchange plus gaming and amusement, but it would be misleading to suggest that there is likely to be a silver bullet to stop cyber-crime and cyber-terrorism. A single answer—a cyber-panacea—will not magically appear to universally protect your financial transactions and to avert all cyber-terrorist attacks. Thus you must be vigilant to undertake the steps outlined in this book and support governmental and corporate protective actions that provide at least a reasonable level of security.

If you operate a business it would be very prudent to follow the five-step program in the new U. S. cybersecurity framework. This means taking action to: "Identify, Protect, Detect, Respond and Recover." The digital technology and the nature of cyber-attacks will undoubtedly change, but this cybersecurity framework will likely remain consistent, effective, and flexibly efficient to respond to cyber-attacks of all kinds. The one thing that will likely change is that the detection process will continue to ramp up, and the penalties for cyber-crimes and especially cyber-terrorists will increase to the highest levels. In short cyber-terrorists, in particular, will be sentenced to life or even the death penalty. And even cyber criminals, including those of a young age, may find themselves sentenced to very long terms in high security prisons. In short the future will continue to spawn cyber-attacks of all types, and vigilance to Identify, Protect, Detect, Respond, and Recover will remain the watchword in this field for a long time to come.

# References

1. "Solar Storms Threatened U.S." May 22, 2013, Lloyds of London. http://www. lloyds.com/news-and-insight/news-and-features/environment/environment-2013/ us-east-coast-at-high-risk-from-solar-storms.
2. Joseph N. Pelton and Firooz Allahdadi (editors), Introduction, Chapter 1, *Handbook of Cosmic Hazards and Planetary Defense* (2015). Springer Press, New York.
3. Richard Stiennon, "Should We Abandon Digital Certificates, Or Learn to Use Them Effectively?", *Forbes*, May 14, 2013. http://www.forbes.com/sites/ richardstiennon/2013/05/14/should-we-abandon-digital-certificates-or-learn- to-use-them-effectively/.
4. Max Anderson, "Bank-Hacking Ring Many Have Stolen $1 Billion," *Washington Post*, February 16, 2015, P. A3.
5. Cliff Skolnick, Wireless LANS Security Issues and Solutions. http://www.toaster. net/wireless/Talks/BAWUG/hotspot_security.pdf.
6. "Top five cloud computing security issues," ComputerWeekly.com, April 24, 2009. http://www.computerweekly.com/news/2240089111/Top-five-cloud-computing- security-issues.
7. *Ibid.*
8. Framework for Improving Critical Infrastructure Cybersecurity, Version 1.0, National Institute of Standards and Technology, February 12, 2014. http://www. nist.gov/cyberframework/upload/cybersecurity-framework-021214.pdf.
9. Cyber Security: Defending SCADA and Industrial Control Systems, Los Alamos Technical Associates, White Paper No. 2, Global Institute for Security and Training, 2014, p. 12.
10. Reuters, "U.S. To Create New Cybersecurity Agency: Official," February 10, 2015. http://www.huffingtonpost.com/2015/02/10/us-cybersecurity-agency_n_6651688. html.
11. Framework for Improving Critical Infrastructure Cybersecurity, Version 1.0, National Institute of Standards and Technology, February 12, 2014. http://www. nist.gov/cyberframework/upload/cybersecurity-framework-021214.pdf.
12. Smart Grid Homepage – National Institute of Standards and Technology, www. nist.gov/smartgrid.
13. Ryan Naraine, "Firefox tops list of 12 most vulnerable apps," *Zero Day*, December 15, 2008. http://www.zdnet.com/article/firefox-tops-list-of-12-most-vulnerable-apps/#!.
14. Apple Invents Facial Recognition Locking & Unlocking System, September 20, 2012. http://www.patentlyapple.com/patently-apple/2012/09/apple-invents- facial-recognition-locking-unlocking-system.html.
15. Gary Mazo, "How to set up Face Unlock on your Android phone," July 14, 2012. http://www.androidcentral.com/how-set-face-unlock-your-htc-one-x-or-evo-4g-lte.
16. Debra Donston-Miller, "The Internet Of Things Poses New Security Challenges," February 25, 2014. http://www.forbes.com/sites/sungardas/2014/02/25/ the-internet-of-things-poses-new-security-challenges/.
17. Jessi Hempel, "Microsoft in the Age of Satya Nadella," *Wired*, February 2015. http://www.wired.com/2015/01/microsoft-nadella/.

# 8

# Ten Essential Rules
# for Your Cyber Defense

The opportunities opened up by the Internet, the world wide web, e-commerce, and the ever-expanding reach of cyberspace are enormous. Just a few of these benefits have include tele-education, tele-health and tele-medicine, e-mail, on-line trading, telebanking, scientific collaboration and research, improved and more cost-effective governmental services, plus social networking with friends and colleagues around the world. This expanding electronic networking allows more and more communications and information exchange, bit obviously entails additional risks. These risks include loss of personal security, exposure to identity theft and spyware, and a variety of cyber-crimes. Anyone living the modern electronic world needs to take precautionary steps to defend against theft, loss of privacy, and possible exposure to criminal charges as perpetrated by someone that wishes you ill. Here are ten practical steps that we recommend that you take to protect your cyber security—not only for yourself but also your loved ones.

## #1. Protect Your Personal Records and Your Passwords

It is important for you to protect your personal records and pass words. The complicated world that we live in today is difficult to handle easily. The proliferation of accounts and passwords are increasingly difficult to keep track of and each account requires a password. It is not uncommon to have passwords for your electronic records at your bank, credit card accounts, utility accounts, brokerage firm, and insurance accounts. This is just for starters. You could also have passwords for pizza companies, Groupon,

© Springer International Publishing Switzerland 2015
J.N. Pelton, I.B. Singh, *Digital Defense*, DOI 10.1007/978-3-319-19953-5_8

Twitter and Instagram, perhaps eight just for airlines and another eight for hotels, service stations or department stores. Each of the authors for this book have more than 30 accounts requiring passwords. Many of these accounts have special rules. This might be for eight or more numbers, letters or symbols. Some require a mixture of letters, numbers and symbols and combination of upper and lower case letters. It is a true challenge to keep track of this unending stream of passwords and not make the mistake of trying to use the same password for everything or use your telephone number, address, birthday or social security number as your one personal code. It is clearly a problem with no easy solution, although new facial recognition systems that substitutes your face for a password, seems to be a promising new approach to password protection. It is important to create a secure hidden file that is comprehensive and provides you easy access to your passwords. There will remain dozens of sites from Social Security to pizza delivery sites that don't offer a facial recognition option.

You should divide your passwords into two categories—vital and non-vital. The first category of vital passwords represent those codes that provide access to accounts of significant value. These are for bank accounts, credit cards, IRA accounts, social security accounts, and those of financial consequence. These should be well protected and not easily detected as to what they are or where they are stored. These should, for instance, be well hidden in a computer file or much better a flash drive with a non-obvious file name. These should never be communicated to anyone on line. If you get an e-mail from your bank or stock broker or from Pal Pay asking for you to update your password, DON'T DO IT! It is a phishing or pharming scam.

Password or PINs (personal identification numbers) for pizza orders, groupons, airline frequent flyer and hotel programs, an AARP or AAA account, or other activities that do not expose you to any major financial loss can be less rigorously protected and you can repeat some of these types of passwords as long as they have nothing in common with a vital password. These can be kept hidden in a desk or file drawer or on-line as passwords. You could be devious and include entirely bogus passwords for vital passwords with your non vital information. A key logger that finds a way into your non-vital password file will think they have scored big. Only later they will find they have been duped. The bottom line is that password protection is vital. Protect your vital passwords.

## #2. Obtain a Secure Place to Store Vital Information

You should also make sure that you find a good way to store vital information such as birth certificate, marriage license, passport, social security documents, bank account and stock broker account information, real estate property and car deeds and will in a safe place such as a bank deposit box. Documents such as a will and testament can be officially filed with local governmental authorities such as a "Clerk of the Courts". Stocks should be held with stock brokers.

In the case of all your vital records—and especially a passport and a driver's license—it is important to keep a Xerox copy of these in your desk, fireproof container, or your safe. Keep a Xerox of your passport and travel documents and driver's license with you when on travel. One of the authors of this book had his passport stolen in the Frankfort, Germany Airport when on travel. Since he had a Xerox copy of his passport information with him he was able to go to the local US Field Office for Citizenship and Immigration and obtain a replacement passport within 1 h.

The problem is that if you have a couple of bank accounts, several credit cards, an IRA account, you could find yourself inundated with paper work. You could go to paperless files, but if there is a major cyber-attack on your bank you might have no official paper records of your assets. The option is to make sure that you use a paper shredder to destroy your records or print out key summaries of your accounts on a once a week basis with some discipline. Some people feel comfortable with all electronic records. People who get detailed paper records actually end up feeling more secure. More importantly they pay more attention to market trends, account irregularities, and problems involving their accounts. The main thing is keep track of vital records and have a secure place to store them.

## #3. Shred All Financial, Medical, and Other Personal Information that You Discard

There was once a simpler time when people threw out there trash with no concern to its contents. They may even have kept their doors unlocked. In a world of identity theft, bogus credit cards, and rampant cybercrime, such carefree actions are unwise and potentially disastrous. There are a number of ways a cybercrook might pick you for a target. One way is to engage in dumpster diving at businesses or even to go through a neighborhood collecting trash bags to see what riches they can find in terms of bank statements, stock broker reports, medical records, social security or driver's license numbers, or even discarded notes with passwords or other personal information. This might be step one in an attempt to either extort funds from you—by seeking money not to disclose embarrassing information—or stealing your identity and your financial assets. Just one discarded credit card or banking statement that includes account numbers, social security number, and your address might be enough for a cyber-crook to begin the process to steal you identity.

In today's world a paper shredder can be your friend. On the other hand, tossing personal and financial information out with the recycling or the trash may reveal vital information as step one in identity theft. This can ultimately lead to illicit electronic transfers of your money and worse. There are shredders that can be purchased for under a $100 that provide reasonable protection and can do multiple pages at a time. Another option is to keep a shredding file and take it to your bank or your local government's free shredding service. This is convenient, especially if this is a trip that you would make in any event. When your records are shredded with many others files by a professional heavy duty shredder and the bagged shredded paper is carted off to a secure site, you have an extra layer of security.

# #4. Get AntiVirus Software Protection

It is essential to obtain an antivirus software package and install it on not only your desktop computer but also on your laptop, tablet and even smart phone. Windows-based software can be vulnerable to intrusions although this has improved since the Windows XP 2001 software was introduced. Any device that you use for financial transactions, paying your taxes, or storing key medical information or records should be protected.

In Chap. 5 we covered the various antivirus software protection packages that you can easily obtain and installed on line. The 12 brand name antiviruses that are listed in Chap. 5 and are offered for a modest annual fee, plus the additional two free systems cited, all provide at least a minimum of protection. These antivirus programs are reasonably quickly updated and thus should let your electronic devices be largely secure unless you are targeted by a highly skilled "cracker."

These various antivirus packages may or may not protect you against spyware and keyloggers that are not trying to infect your computer, but rather they are trying to steal your identity and thus trying to steal your passwords, account information and financial assets. Antivirus systems are designed to protect your computer and electronic devices against being infected, but do not necessarily protect your financial assets. Firewall protection is thus important to protect against those using spyware programs and keylogging that can capture your keystrokes. This we topic will return to in "Commandment No. 6."

## #5. Prevent Your Identity from Being Stolen

You actually need to devote some effort to prevent your identity from being stolen. The ways that identities can be stolen and the tools that can be used to accomplish this cyber-theft just keep on increasing. The reasons for identity theft and how this can be done are also diverse. The chart provided in Fig. 8.1 has been prepared by the U. S. government to show this diversity. The motivations break down as follows: Credit-related fraud accounts for 20 %; Bank loan fraud accounts for 13 %; Phone and utilities fraud accounts for 13 %; Various types of employment fraud and related misrepresentation accounts for 15 %; government document frauds such as fraudulent claims for social security, Medicare/Medicaid accounts for 15 %; a wide range of other scams and financial frauds accounts for the remaining 24 %. Overwhelming, the purpose is to get money or get out of paying one's bills and shifting the responsibility to others whose identity has been stolen. Credit card fraud, fraudulent loans, and bogus payments for utilities dominate the reasons that identities are stolen. There are more serious reasons

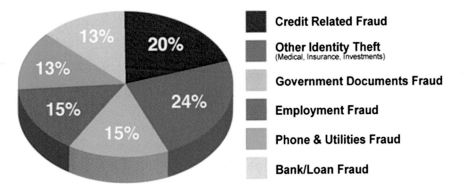

Source: www.ftc.gov

**Fig. 8.1** Various types of identity theft activities. (Graphic courtesy of the U. S. Federal Trade Commission, www.ftc.gov.)

COMPLAINTS

Top Consumer Complaint Categories*

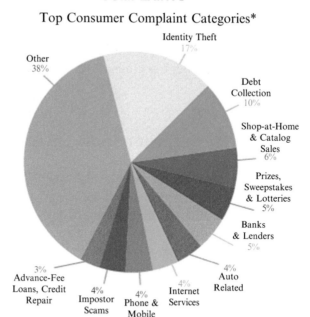

**Fig. 8.2** Complaints from consumers to the FTC. (Graphic courtesy of the Federal Trade Commission, www.ftc.gov.)

related to medical records, extortion, and even framing intended victims for crimes they did not commit for revenge or for some time of extortion.

To put the threat identity theft into perspective it is the number one reason that consumers complain to the U. S. government about fraudulent or illegal behavior. Again the other chart from the Federal Trade Commission shows the nature of complaints received (Fig. 8.2).

Steps that you should take include the following: (a) Do not take out too many credit or debit cards and cancel those that you do not use; (b) Monitor your credit card accounts for any unusual charges; (c) check your credit score with the Equifax, Transunion, Experian credit reporting services; (d) protect your credit cards, bank accounts, stock or real estate holdings and reports about them; (e) use at least eight digit passwords for these account and ensure that they have a strong protective coding (i.e. upper and lower cases of letter, symbols and numbers) and never send these codes to others electronically. Also update at least once a year and never be duped by an e-mail or attachment form that asks for your password; and (f) Get a basic firewall protection plus an identity protection service and warranty. If you have assets above $50K it is worth in the range of $200–$300 a year it takes to get an antivirus, firewall and ID protection policy.

# #6. Obtain at Least the Basic Personal Firewall Protection

The need for firewall protection for corporations, especially large global enterprise networks with huge internal intranets, is clear cut. Such corporations would be subject to legal suits if they did not have such protections in place and there was a large breach of their intellectual property or financial holdings. Even for individuals, a firewall makes sense. If there is a simply a wireless router or family Local Area Network with a mother, a father, three children and perhaps a grandmother accessing the Internet through one node, a firewall makes a good deal of sense. Even one or two uses in a household might find a firewall of value. In light of the pricing policies by many vendors it is often possible to get both antivirus and personal firewall protection for a consolidated annual price that can be as little as $35–$60 a year. This protection—together with an identity theft protection and $1 million guarantee or warranty is something that makes sense in today's world so rife with cyber-criminal protection.

One addition step is important to take. If you have a wireless router that is serving your entire household, you should make sure that access to your router is protected. Without password protection your neighbors or someone driving through your neighborhood with a scanner could intercept your messages, log on to the internet and distribute spam or worse and it would seem to come from you. Unprotected wireless routers in household with unprotected passwords is one of the biggest dangers today. Cybercrooks can obtain a dangerous amount of information in this manner.

# #7. Recognize Warning Signs and Respond Quickly to Threats

It is easy to be oblivious to cyber-attacks. When someone robs you at gun point or breaks into your home the commission of a crime is abundantly clear. Some cyber-criminals can be quite stealthy. Some particularly clever "crackers" or "black hat" hackers have found ways to create Trojan horse accesses into banks and then over time found ways to withdraw a *de minimus* amount from all the accounts each day without detection for a long period of time. Crafty electronic crimes can be hard to detect. One should certainly make sure that your credit card has not been compromised by looking at the statement and not having many cards active. If you lose a card, immediately report it missing or stolen (as the case may be) and

Table 8.1 Example of more sophisticated phishing e-mail seeking you to open attachment

| Department of the Treasury | |
|---|---|
| Internal Revenue Service | |
| *Tax year ended* | *December 31, 2013 to December 31, 2014* |
| Deficiency | $11,995.15 |
| Increase in tax | $1312.54 |
| Penalties or additions to tax | $830.12 |
| IRC 6662(a) | |
| The Deficiency Statement and Contestation Form are attached to this email. Please review the attached documents! | |
| We are at liberty to seize your bank account(s), effective immediately. Withdrawal of any amount, prior to addressing the attached document constitutes a violation of the law. | |
| NOTICE OF DEFICIENCY | |
| We have determined that you owe additional tax or other amounts, or both, for the tax year(s) identified above. This email is your NOTICE OF DEFICIENCY, as required by the law. The attached statement shows how we figured the deficiency. | |
| If you want to contest this determination in court, you have 3 days from the date of this email (15 days if this email is addressed to you outside of the United States) to file a petition with the United States Tax Court for a redetermination of the deficiency. You can get a copy of the rules for filling a petition and a petition form you can use by writing to the address below: United States Tax Court, 400 Second Street, NW, Washington, DC, 20217 | |

arrange for a replacement. If you have real estate holding make sure that liens from illicitly obtained loans have not be placed against your holdings.

The warning signs may not always be obvious as they once were when you got a solicitation from a Nigerian Prince out of the blue. You might today get a message that appears to be from a legitimate address such as from "paypal.com" or even "irs.gov". These legitimate appearing e-mails, however, can be generated through cyber tricks. If you were to get such a solicitation as in Table 8.1 below via mail, it would be a serious mail fraud felony. But in the cyber world this is just another spam solicitation. If an e-mail with an attachment looks in any way bogus don't open it. Even if comes from a friend on known e-mail address, check first with your friend or associate. Your friend's e-mail may have been captured by a netbot that is forwarding a virus or some other form of malware.

# #8. Consider Getting Comprehensive Insurance Against Financial and Cyber-Loss

The further step that you might consider goes beyond getting an antivirus, firewall, identity theft coverage is to go to one of the major insurance carriers and sign up for a comprehensive policy that will provide you everything discussed in this short book and more. This will provide you with all possible forms of coverage and more things than you would ever think of yourself. In Chap. 5 we discussed the type of coverage that Met Life Defender offers as an example of all of things that are covered. You could likely even negotiate more than $1,000,000 in warranty coverage. The bottom line is that the various individual pieces of cyber-protection will likely cost you less and are relatively easy to accomplish yourself. If we wish to be thorough you might check out this option. Such coverage is much more often obtained for corporations and those that have a fiduciary responsibility to customers such as banks, brokerage firms, accounting firms and law firms. There are a lot of ins and outs to such policies. Corporations must address whether there is third party coverage, whether coverage is for encrypted and non-encrypted cyber-services, etc. For corporations considering such services it is recommended that you go to the following site: Top Ten Tips for Companies Buying Cyber Security Insurance http://www.acc.com/legalresources/publications/topten/tttfcbcsic.cfm

# #9. Support the Adoption of Stricter Governmental Legislation and Standards for Cybersecurity

There is a great deal more that can be and should be done to protect consumers and indeed national populations against cyber-attacks and potentially cyber terrorism. If someone sends you a fraudulent offer by mail they are guilty of federal mail fraud that is a serious felony. If someone comes to your door and make fraudulent offers they too can be prosecuted. The Internet is still very much the "Wild West" that on one hand allows innovation, but on the other invites fraud and criminal activity, just because it is so hard to be found and prosecuted—especially if operating from Eastern Europe or Africa or some other locale where there is no extradition policies in effect. One reform could be that fraudulent offers and cyber-attacks would be the legal equivalent of mail fraud.

Beyond the lack of strong legal enforcement powers against cyber-criminals committing crimes on-line, there is the even more serious issue of Level 2 attacks against corporations and Level 3 attacks that are against a national and its vital infrastructure. A member of the British Parliament MP James Arbuthnot recently spoke out about cyber-attacks on the United Kingdom's the energy sector—which is overseen and protected by what is call the National Grid. He revealed that currently there is about once a minute an attempted cyber-attack on the electric power grid. One can assume that the attacks on other advanced economies, including the United States, are at a comparable if not higher level. In today's advanced economies, where electrical power is central to transportation systems, virtually all forms of employment, medical and educational services, a successful cyber-attack on electrical grids would be devastating. The same is true with regard to other vital infrastructure such as water supply, sewage systems, pipeline systems, etc. Protecting a nation against cyber-attacks today is core to national defense. Everything from nuclear power plants to air lines, from rail systems to water and sewage systems, and from the electric power grid to elevator and escalator systems are today at risk via cyber-attack. Stricter national legislation and stricter standards for industry protection are core to one's own safety and welfare in a cybernetic world. In short we need greater cybersecurity at all levels—Level 1-personal cybersecurity; Level 2-Industry cyber-security; and Level 3-National cybersecurity.

Serious attempts to upgrade the security of the Internet should be carefully considered and if the certificate authentication process can be truly verified such changes should also be supported.

## #10. Take Common Sense Steps to Guard Against Cyber-Attacks

Every family and every home in the United States and around the world should take steps toward greater cyber-security. The first step is to protect all of your electronic devices that connect to the Internet and any IT system. The big three are antivirus, a firewall, and identity theft protective services. Additional insurance protection is possible. You should systematically shred personal, medical and financial documents, undertake systematic protection of your electronic passwords and ensure that these are of sufficient rigor and complexity to not be easily guessed or decoded. Do not open messages from unknown or suspect sites and be particularly attuned to not open attachments from spam messages. Even messages from friends that seem suspect should be checked by calling them or e-mailing to see if this is something they sent or perhaps it came via a netbot. It may sound trite, but if you have an "Internet native" (that is someone who grew up always using computers and familiar with Internet trends and terminology) close at hand, then check with them first if something seems wrong or suspicious.

Another common sense protection is to be caution of which applications you chose to use on your smart phone—especially any that involve banking or credit card activities. Recently it was reported that nearly 6 % of android applications that were built using Apache Cordova platforms are vulnerable to being compromised. It was further reported that of the nearly 250 applications that involved banking that 25 of them or 10 % were vulnerable to attack by a cybercriminal. In short millions of users who use Android Cordova-based apps, that have not been updated to the latest version, are at risk of having sensitive information, such as their login credentials, stolen from these applications. This could potentially allow attackers to impersonate them, access their accounts and even make purchases on their behalf. The updated version 3.5.1 of Apache Cordova is now protected due new software updates, but this is but one example. Other smart phone apps will be compromised. I-Phone and Android phone use is convenient, but there are security risks [1].

Social Media is also something to be cautious of as well. In June of 2013 Facebook revealed that it had inadvertently revealed to black hat hackers contact information for over six million of its users. This is a much less serious problem than a vulnerable banking application, but who knows what use hackers might make of your contact information when linked to all the information that you may have revealed on your Facebook page [2].

You do not have to become a cyber-expert and know every security breach that occurs, but caution is in order. Some technical savvy is useful. It is probably useful to get to know and understand the terms provided in the glossary and acronym section of this book. Most of all stay alert and aware.

# References

1. HAY Roee Hay "IBM X-Force Finds Apache Cordova Vulnerability That Might Expose Nearly 5.8% of Android Apps", Security Intelligence.com August 5, 2014. http://securityintelligence.com/apache-cordova-phonegap-vulnerability-android-banking-apps/#.VPJv5PnF_OE
2. Alex Fitzpatrick, "Facebook Exposed 6 Million Users' Contact Info" August 2013, Mashable.com. http://mashable.com/2013/06/21/facebook-data-breach/.

# Appendix A
# Glossary of Definitions
# and Acronyms

**Adware**   These are ads that appear on your computer or phone screen uninvited. They are usually enabled by installing freeware or shareware on your computer.

**Analog coding or scrambling**   This is a way of coding radio or television signals by distorting the signal so that an analog descrambler can restore it. This system is not very secure, and analog descramblers can be purchased that receive scrambled TV signals. (See **Digital encoding**).

**Android operating systems**   These operating systems are issued by Samsung and are the basis for developing applications for Android phones. They can reveal someone's user ID and potentially lead to other security breaches.

**Antivirus**   Antivirus software (often abbreviated as AV), sometimes known as anti-malware, is computer software used to prevent, detect, and remove malicious software. Antivirus software was originally developed to detect and remove computer viruses, hence the name.

**AP**   Access point in a wireless local area network (WLAN).

**API**   In computer programming this refers to an application programming interface (API). Such an interface computer standard is a set of routines, protocols, and tools for building software applications. This is key in the use of "The Cloud" and, in particular in terms of whether the use is as an "Infrastructure as a Service (IaaS)," as a "Platform as a Services (PaaS)," or as a "Software as a Service (SaaS)."

**APT**   Advanced persistent threat. These are threats posed by techno-terrorists or sophisticated cyber criminals. APT threats are the focus of the U. S. Cyber Command and other national attempts to defend against the most sophisticated of black hat hackers.

© Springer International Publishing Switzerland 2015
J.N. Pelton, I.B. Singh, *Digital Defense*, DOI 10.1007/978-3-319-19953-5

**Back doors**    These are ways to use the root-level machine instructions to access a computer. This task can be accomplished by rootkits or in other modes of attack via Trojan horse malware. Malware is often enabled by an attacker's ability to bypass normal authentication to gain access to a computer, electronic tablet, or smart phone.

**Backup memory**    One key form of protection of one's data files is to automatically back them up on zip memory sticks or have a protective service that automatically backs up ones files. Most corporations and governmental agencies have their files backed up at an off-site location to protect key files.

**Black hat hacker**    This is the term that is given to a cracker or someone who violates computer security for illegal gain or other nefarious or even terrorist purposes.

**Blog**    This is the putting together of "web" and "log" to form a blog. It refers to any ongoing posting on the web of general information, news, or personal information for anyone to access.

**Bootkit**    This is a more sophisticated form of a rootkit. More specifically it is a "kernel-mode" variant of a rootkit. It can be used to attack computers that are protected by full-disc encryption.

**Bot**    This is a targeted computer or targeted processor such as a device installed in an appliance (see **Internet of Things**) that is taken over by a so-called bot-herder. Once a targeted machine is taken over by malware, the computer or processor can become a part of a botnet (also known as a zombie) that sends out spam or be used to engage in phishing, etc.

**Bot-herder**    This is a slang term that refers to a black hat hacker that controls a botnet.

**Botnet**    This can be as mundane as keeping control of an Internet Relay Chat (IRC) channel, or it could be used to send spam e-mail or participate in distributed denial-of-service attacks. The word botnet is a combination of the words robot and network. The computers that form a botnet can be programmed to redirect transmissions to a specific computer, such as a website that can be closed down by having to handle too much traffic—a distributed denial-of-service (DDoS) attack—or, in the case of spam distribution, send a message to many computers.

**CDU**    Cyber Defense Unit of Japan.

**CIO**    Chief Information Office or Chief Information Officer.

**CISO**    Chief Information Security Office or Chief Information Security Officer.

**Click fraud**    Click fraud is the method of generating inflated numbers as to traffic on a commercial website. This is particularly the case where ad viewings are tied to payments for online ads. The fraudulent viewings are either using non-human sources—such as lines of code that automatically click on brands' ads—or hiring a number of users to manually click on the same ads in order to increase the amount of revenues tied to ad viewings.

**Clone phishing**     This is one of the more sophisticated and effective forms of phishing. In this case the attacker will first hack into one of your trusted contact's e-mail account and then resend a previously sent e-mail from the trusted e-mail address. However, the attacker will have first modified key aspects of the original e-mail, replacing a legitimate link, reply to address, and/or attachment with a harmful one. Even if you yourself use strong passwords and security for your e-mail, of all the many contacts you have, it is likely that not all of your acquaintances take such strong precautions, leaving not only them but you vulnerable as well. This means it is important to be cautious and exercise good judgment when clicking on any link or attachment regardless of whether the source of the e-mail appears to be trustworthy.

**Cookie**     Cookies are small files that are stored on a user's computer. These files are designed to hold a modest amount of data that is specific to a particular computer and website. The "cookie" or file can be accessed either by the web server or the client computer. This allows the server to deliver a page tailored to a particular user. In order to find out whether your browser allows your "cookie" to be captured you need to go to the "cookie checker."

**Cracker**     A cracker (also known as a black hat hacker) is an individual with extensive computer knowledge whose purpose is to breach or bypass Internet security or gain access to software without paying royalties. The general view is that, while hackers build things, crackers break things. Cracker is the name given to hackers who break into computers for criminal gain, whereas hackers can also be Internet security experts hired to find vulnerabilities in systems.

**Cross-site scripting**     This activity is similar to website forgery. Cross-site scripting injects a malicious script into a victim's computer so that when a user accesses a legitimate trusted site and submits his personal information, the hacker is able to intercept the transmission to steal the information. Software and scanning services are designed to protect against such attacks, but this is a constantly evolving battle between the cyber criminals and the cybersecurity professionals, so it is important to keep your Internet-connected devices up-to-date with the latest protective tools.

**Cyber-attack**   There are number of different strategies for a so-called stealth cyber-attack on a network user. These strategies include: (1) **Detection Evasion**: This type of attack—the most common—seeks to evade the security system used on your network and individual computer. The attacker moves the root level and bypasses the operating system in seeking to avoid the anti-malware and other security software on your network. (2) **Targeting**: This type of attack is targeted at a particular organization's network. It creates an attack website through which many individuals can attack another specific site. (3) **Dormancy**: The attacker plants a malware (Trojan horse or time bomb) and then waits for a later time to mount an attack. (4) **Persistency**: The attacker keeps on trying until he or she gets access to the network. (5) **Attack Cover through Complexity**: This method involves the creation of noise as a cover for malware to enter the network.

**Cybersecurity**   Methods and tools that can be used to protect one's online privacy and to prevent digital attacks on one's computer, smart phone, or other electronic devices.

**Dark Net**   An encrypted website that allows anonymous access. It was established to allow citizens in dictatorships to be able to communicate openly, but it is today being used by cyber criminals and techno-terrorists.

**DARPA**   Defense Advanced Research Projects Agency.

**Data bomb (or logic bomb)**   A logic bomb is a piece of code intentionally inserted into a software system that will set off a malicious function when specified conditions are met. For example, a programmer may hide a piece of code that starts deleting files (such as a salary database trigger) should they ever be terminated from the company. Software that is inherently malicious, such as viruses and worms, often contain logic bombs that execute a certain payload at a pre-defined time or when some other condition is met.

**DDNS**   Dynamic DNS is a technique used to update a domain name system (DNS) server record for networked devices in real time.

**DNS**   Domain name system. (See **Phishing** and **Pharming** for how this system can be abused by a black hat hacker.)

**DDoS**   This a distributed denial-of-service (DDoS) attack. In this case a network of botnets are formed using malware and then totally overload a website with traffic so that it is not able to operate in a normal mode.

**DHS**   Department of Homeland Security, the division of the federal government charged with our country's security.

**Digital encoding (encryption)**   This is a much more secure form of encoding a signal that would require a computer processor a considerable time to ever decode. If the key to decode a digital signal were sufficiently complex, such as a 62-bit or 128-bit code, then it would be virtually impossible to decode.

**Disabling audit function**   Disabling computer audit functions in order to disguise the presence of viruses, malware, or Trojan horse time bombs

**DOD**   Department of Defense of the United States.

**Domain name system (DNS) server**   In computer networking, the Domain Name System (DNS) is a hierarchical distributed naming system for computers, services, or any resource connected to the Internet or a private network using the Internet Protocol. The key function of the DNS server is to translate a specific domain name assigned to each of the participating entities. Most prominently, it translates domain names, which can be easily memorized by humans, to the numerical IP addresses needed for the purpose of computer services and devices worldwide. The domain name system is an essential component of the functionality of most Internet services.

**Drive-by-download**   This is computer hacker slang that refers to things like adware that is secretly downloaded along with freeware such as free greeting card services. Such malware is thus downloaded onto a computer from the Internet without the user's knowledge or permission. Most "freeware" on the Internet thus comes with a cost.

**EINSTEIN 3**   This approach, now known as EINSTEIN 3, draws on commercial information technology and specialized government technology to protect the U. S. government's data networks. The system conducts real-time full packet inspection and threat-based decision-making on network traffic entering or leaving Executive Branch networks. Einstein 3 is deploying and testing intrusion prevention systems across the federal arena. This Initiative represents the next evolution of protection for civilian departments and agencies of the federal executive branch. The goal of EINSTEIN 3 is to identify and characterize malicious network traffic, to enhance cybersecurity analysis, perform situational awareness, and implement appropriate security response. It is to automatically detect and respond appropriately to cyber threats before harm is done by creating an intrusion prevention system supporting dynamic defense. EINSTEIN 3 will assist the Department of Homeland Security (DHS) in defending, protecting and reducing vulnerabilities on federal executive branch networks and systems. The EINSTEIN 3 system will also support enhanced information sharing by US-CERT with federal departments and agencies by giving DHS the ability to automate the alerting of detected network intrusion attempts and, when deemed necessary by the DHS, to send alerts to the National Security Agency (NSA).

**Electronic village**    The concept of the world being closely connected via electronic media as first presented by Prof. Marshall McLuhan in his writings.

**EMV chip**    This is a chip inserted into a credit card that prevents counterfeiting, which is much easier if your credit card simply has a magnetic strip with your credit card data embedded in it. EMV stands for "Europay, MasterCard and Visa," that together developed this chip.

**Encryption**    This means to code information to protect it being read by or accessed by anyone except the intended reader. Decryption is the process of decoding the message with a decryption key so that it can be read.

**ENISA**    The European Union Agency for Network and Information Security, which is headquartered in Crete in Greece; its mission and activities are described in Appendix D of this book.

**FBI**    Federal Bureau of Investigation.

**Firewall**    This is a network security system that controls the incoming and outgoing network traffic in order to protect the internal network against spam and malware such as worms, viruses, zip and logic bombs, and Trojan horses. A firewall is based on a set of rules that isolates and protects the internal network from harmful software and cyber-attacks. A firewall thus establishes a barrier between a trusted, secure internal network and another network (e.g., typically the Internet) that is assumed not to be secure and potentially be the source of harmful malware. A firewall can be used within a home-based network, a small office, or an entire corporate enterprise network.

**FISMA**    The Federal Information Security Management Act of 2002.

**Global Brain or the World Wide Mind**    Concept of how advanced electronic networking and artificial intelligence could link human intelligence together in ever closer ways to speed up thought processes and technological innovations. This rate of technical advancement, however, could contain a wide range of political, economic, social, and cultural hazards.

**GUI**    Graphical User Interface, a GUI (pronounced as either G-U-I or gooey) allows the use of icons or other visual indicators to interact with electronic devices, rather than using only text via the command line. A GUI uses windows, icons, and menus to carry out commands, such as opening, deleting, and moving files. One of the vulnerabilities of mobile devices is related to a graphical user interface that could be tricked into hiding a security dialog.

**Hacker**    This term refers to those with extensive computer science or system analysis background. So-called black hat hackers or crackers use their knowledge and skills to break into computer networks for illegal gain or other purposes against the interests of the user community.

**Hacktivists**   This term applies to those that hack into computer networks to obtain and reveal information that they feel is being withheld from the public and that it is worth the risk of undertaking a network incursion to reveal this information. Wikileaks is perhaps the prime example of what might be called hacktivism. Edward Snowden, who is under U. S. federal indictment for revealing top secret information, has claimed that this was the purpose of his activities.

**HTTP**   Hypertext is structured text widely used on the Internet and the World Wide Web. HTTP uses logical links called hyperlinks to connect from one node on the Web to another. Thus HTTP is the protocol to exchange or transfer hypertext. The standards development of HTTP was coordinated by the Internet Engineering Task Force (IETF) and the World Wide Web Consortium (W3C). The standard for HTTP was defined by IETF Requests for Comments (RFCs), especially in RFC 2616 (June 1999), which defined HTTP/1.1, the version of HTTP most commonly used today. In June 2014, RFC 2616 was retired and HTTP/1.1 was redefined by a new series of Request for Comments known as RFCs 7230, 7231, 7232, 7233, 7234, and 7235.

**HTTPS**   Hypertext Transfer Protocol Secure (HTTPS) is a widely used communications protocol for secure communication over a computer network, with especially wide deployment on the Internet. Technically, it is not a protocol in itself; rather, it is the result of simply layering the Hypertext Transfer Protocol (HTTP) on top of the SSL/TLS protocol, thus adding the security capabilities of SSL/TLS to standard HTTP communications.

**IaaS**   Infrastructure as a Service. This is a term used in the provision of various types of services from The Cloud. Infrastructure as a service (IaaS) typically offers a choice of open Cloud infrastructure services for various types of information technology operations. There are often fully managed IaaS types of offerings that can be used to develop applications and run production-ready workloads, or so-called "soft layer" IaaS offerings that involve less expense and a lower level of support. (See also **PaaS**, or Platform as a Service, and **SaaS**, or Software as a Service).

**Identity Theft**   This is the stealing of one's identity usually for the purpose of illegal gain. Identity theft can be accomplished through physical means such as robbery or obtaining discarded records. In today's cyber world, the usual means of identity theft is through hacking into someone's computer and obtaining personal identifiers such as Social Security numbers, banking and brokerage accounts, and associated access codes or personal identification numbers. Identity theft could be used to create an alternative identity under which name a crime or even act of terrorism might be conducted. Thus protection of a personal identity is crucial in a world in which financial and business activities are increasingly electronic.

**IEEE**   Institution of Electronics and Electrical Engineers. This large professional organization also develops standards. Its standard for Wi-Fi wireless networks, namely 802.11 in its various forms, is key to the provision of wireless access services.

**Internet of Things**   The latest trend is to make all sorts of appliances, machines, and electronic devices (such as refrigerators, washing machines, security systems, automobiles, boats, buses, etc.) "smart" to the extent that they contain digital processors and the ability to communicate via the Internet. This means that so-called botnets that can engage in such activities as distributed denial of service (DDoS) can in the world of the Internet of Things come from literally billions of these "smart" devices.

**iOS**   This is the operating system for the I-Phone, the I-Pad, and all the other Apple devices including the Apple Watch. Developers of applications for the I-Phone and I-Pad often use kits that can contain security leaks, leading to potential attacks on credit cards, according to Appthority and other security reviewers. The same is true for Android applications as well.

**IPSec**   An open-standard Internet protocol used for secure Virtual Private Network (VPN) communications over public IP-based networks. The packet address header that is stripped off by IPSec constitutes a problem for satellite transmission and thus a special interface protocol must be used for satellite transmissions to avoid this problem with IPSec.

**Keylogger**   A keylogger is a type of surveillance software that is typically labeled as spyware. Such spyware has the capability to record every keystroke you make to a log file, usually encrypted. A keylogger recorder can record instant messages, e-mail, and any information you type at any time using your keyboard. This is how passwords to electronic financial records at your bank or stockbroker, Social Security records, etc., can be stolen by a black hat hacker who is intent on stealing your financial assets.

**Knobot**   This is a "knowledgeable robot," or artificially intelligent robot that is independently able to search the web and carry out other functions independently.

**LAN**   Local Area Network.

**Macros**   Shortened name for what is called a "macroinstruction" in computer programming. Specifically, in computer science a macroinstruction is a lengthy rule or pattern that specifies how a certain input sequence (often a sequence of characters) should be mapped to a replacement output sequence (which is also typically a sequence of characters). These detailed procedures represent a location where malware might be implanted. Enabling the insertion of a new macro is something to be done with caution and only when you have an active antivirus program working on your computer or smart phone.

**Malware** Refers to all types of intrusive software that have a malicious intent. Thus included in this group are such things as adware, worms, Trojan horses and time bombs, data bombs, zip bombs, logic bombs, rootkits and bootkits, ransomware, phishing and pharming activities, and more.

**Man-in-the-Middle (MitM) or Rogue Wi-Fi (sometimes known as "Evil Twins")** A MitM or rogue Wi-Fi attack is where a cyber-criminal either sets up a public Wi-Fi hotspot or compromises an existing public Wi-Fi network to attack anyone who accesses it. Sometimes criminals create an "Evil Twin" hotspot located geographically near a legitimate Wi-Fi provider and then give it a nearly identical name to the trustworthy provider. These various Rogue Wi-Fi networks prey on anyone who tries to use Wi-Fi on their smart phones, laptops, tablets, and other Internet-connected devices to access the network, unaware that the criminal has designed it to intercept and/or alter the data that users send and receive. Once connected to such a network, all transmissions become vulnerable to the attacker, who can steal personal information, infect the users' devices with malicious software, or even impersonate trusted contacts. Although software can sometimes protect against such attacks by authenticating a secure connection, prudent users should never connect to a Wi-Fi network that is not known and trusted.

**MBR** Master boot record.

**Memory scraping** This is a malware technique that is often use to defeat point-of-sale security. It is a type of malware that helps hackers to find personal data that is often used in conjunction with credit card validation at point-of-sale verifications and examines memory to search for sensitive data that is not available through other processes. Although data encryption is widely used to secure data, memory scraping finds weak areas from which it can take data. For example, some memory-scraping malware steals encrypted data from applications through which the data passed unencrypted and is thus still potentially accessible. This renders many typical security encryptions vulnerable to attack.

**Network mapper**   A security scanner used to discover network hosts and also identify services provided. It is also sometime referred to as an Nmap.

**Next-generation firewall (NGFW)**   A firewall that provides the latest capabilities that are beyond traditional port-based controls and enforces specifically defined policies that are typically based on application, content, and/or the user.

**NFC**   Near Field Communications. This is the radio frequency ID (RFID) technology that is being used for instant pay and go systems involving credit cards that are registered with banks.

**NIST**   National Institute for Standards and Technology.

**NSA**   National Security Agency.

**PaaS**   Platform as a Service. This relates to companies that utilize The Cloud. Platform-as-a-Service (PaaS) solutions from The Cloud can now be used both to build and deploy new applications—especially for mobile users. Many PaaS providers have extended their offerings to include so-called back-end infrastructure, namely storage and computing, as needed.

**Packet or IP spoofing**   The basic protocol for sending data over the Internet network and many other computer networks is the Internet Protocol ("IP"). The header of each IP packet contains, among other things, the numerical source and destination address of the packet. The source address is normally the address that the packet was sent from. By inserting a fake header so it contains a different address, an attacker can make it appear that the packet was sent by a different machine. The machine that receives spoofed packets will send a response back to the forged source address. This technique is obviously only used when the attacker does not care about the response or has some way of guessing the response. It is sometimes possible for the attacker to see or redirect the response to his or her own machine. The most usual or simplest case in which packet spoofing might be used is when the attacker is spoofing an address on the same local area network (LAN) or wide area network (WAN).

**PCAP**   Packet capture.

**Pharming**   Pharming is an even more devious way of capturing information than "phishing" (see below). Phishing attempts to capture personal information by trying to trick users to visit a fake website. Pharming is an attempt to send users to false websites, but by manipulating the IP website address it can do so without users even being aware that this has happened. Although a typical website uses a domain name for its address, its actual location is determined by its numerical IP address. When a user types a domain name into his or her web browser's address field and hits

ENTER, the domain name is translated into an IP address. This is accomplished by what is called a DNS server. The web browser then connects to the server at this IP address and loads the web page data. After a user visits a certain website, the DNS entry for that site is often stored on the user's computer in a DNS cache. This way, the computer does not have to keep accessing a DNS server whenever the user visits the website. One way that pharming takes place is via an e-mail virus that "poisons" a user's local DNS cache. It does this by modifying the DNS entries, or host files. For example, instead of having the nine-digit IP address 17.254.3.183 direct to www.apple.com, it may direct to another website determined by the hacker. Pharmers can also poison entire DNS servers, which means anyone that uses the affected DNS server will be redirected to the wrong website. (For more information on Pharming go to: http://techterms.com/definition/pharming.)

**Phishing**    This is one of the most common categories of online scams. In this case a criminal, often by means of high volume spam e-mails and/or the establishment of fake websites set up to appear to be legitimate, convinces victims to provide personal information. This might be such data as private account details, credit card numbers, and/or Social Security numbers. If you receive large amounts of unsolicited email and spam in your inbox, chances are that a fair share of these are not simply online businesses looking for customers but instead devious phishing attempts. Basic phishing attacks do not require a high level of sophistication by the criminal and are therefore easy to perpetrate in high volume. Phishing relies on tricking the victims, and while exercising good judgment and online awareness can generally thwart such attacks, many unwary web users still fall victim to such scams every day. Even if you feel you have a keen eye for identifying scams in your e-mail or try to avoid visiting harmful websites, cyber criminals are always working on ways to up their game. It is important to always remain cautious about what websites you visit, how you access them, and who you are providing your personal information to online. Remember that the most effective phishing attempts will always appear on the surface to be legitimate commercial communications or links to bona fide websites. Even though an e-mail may appear to be from your credit card company, remember looks can sometimes be deceiving, and domain names can be tampered with by cyber criminals. There are different types of phishing activities. (Also see **Clone phishing**, **Spear phishing**, and **Whale phishing**).

**PUPs**    Potentially unwanted programs.

**Ransomware**   This is a particular type of Trojan horse that implants malware on a computer so that a special code is needed to unlock a filter that blocks access to all files on that computer. A currently rampant version of a ransomware malware is known as CryptoWall 2.0.

**RDP**   Remote desktop protocol.

**RootKit**   A rootkit is a stealthy type of software, typically malicious, and initially developed for Unix-based systems. It was designed to hide the existence of certain processes or programs from normal methods of detection and enable continued privileged access to a computer. This is what might be called particularly "stealthy" Trojan horse software. A rootkit is able to keep files, registry keys, and network connections, and can keep itself screened from detection. It is called a rootkit because it enables those that use this software to have "root" access to the computer. This means it operates at the lowest instructional level of the machine. A rootkit is designed to intercept common API calls and thus can keep certain files hidden from display, even reporting false file counts and sizes to the user. Rootkits started out as a set of altered utilities for Unix, such as the ls command, which is used to list file names in the directory. It initially had what were considered legitimate uses by manufacturers that wished to monitor computer performance and reliability without imposing on the need of users to respond. This type of backdoor access is no longer considered legitimate. (Also see **Bootkit**.)

**SaaS**   Software as a Service. This is a term that relates to provision of software to users of The Cloud to obtain access to various types of software. See also **PaaS** (Platform-as-a-Service) and **IaaS** (Infrastructure as a Service).

**SCADA**   Supervisory Control and Data Acquisition system that provide the automated 24/7 control and data reporting capabilities for electrical power grids, pipelines, traffic signaling systems, water treatment and distribution, sewage treatment, and other large networks found within "smart cities" and national transportation, communications, and power systems.

**Self-replicating codes**   Certain types of programs are able to self-replicate. They are thus able to spread copies of themselves, and in the most sophisticated forms are able to distribute modified copies. These can be classified as either virus or worms. These types of malware codes have the ability to propagate and distribute themselves to other users' computers.

**SIM**   Subscriber identity module.

**Spear phishing**   This is a form of phishing that more directly targets its victims or specific individuals, either as part of an organization, or because of affiliation or prominent status. Unlike basic phishing attacks, where attackers send out malicious spam en masse, spear phishers know who they are targeting and use that information to their advantage. Spear phishers craft much more convincing e-mails that may appear to be from someone trusted or with authority within an organization. As convincing and personal as an e-mail may seem, it is important to verify that it is authentic and from the correct e-mail address, especially if it includes a request for sensitive information, an unfamiliar link, or a potentially harmful attachment.

**Spy sweepers**   A spy sweeper is a software product that is designed to detect and subsequently assist in removing spyware and viruses from personal computers. This is not a feature automatically installed in antivirus software and typically involves a premium payment for this service.

**Spyware**   Software that can monitor keystrokes and other information and allow hackers to obtain personal identification numbers, access codes, and other personal information.

**SQL**   Structured query language.

**SSH**   SSH refers to what is known as secure shell or sometimes as secure socket shell. There are two versions, known as SSH-1 and SSH-2. SSH represents a cryptographic network protocol for securing data communication. It can create a secure channel by connecting an SSH client application with an SSH server. It was revealed in the so-called "Snowden documents" that the National Security Agency could decrypt such secure shells.

**SSID**   Service set identifier that recognizes those able to access a wireless local area network (LAN).

**SSL**   Secure sockets layer (SSL) technology represents the standard for encrypted client/server network connections. SSL helps to improve on the cyber security of Internet connections. (**Also see HTTPS and SSL/ TLS Protocol**).

**SSL/TLS Protocol**   Perhaps the most popular implementation of public-key encryption is the secure sockets layer (SSL). Originally developed by Netscape, SSL is an Internet security protocol used by Internet browsers and web servers to transmit sensitive information. SSL has become part of an overall security protocol known as transport layer security (TLS). The "S" in HTTPS connotes "SSL." In your browser, you can tell when you are using a secure protocol, such as TLS, in a couple of different ways. You will notice that the "http" in the address line is replaced with "https," and you should see a small padlock in the status bar at the bottom of the browser window. When you're accessing sensitive information, such as an online bank account or a payment transfer service such as "PayPal," chances are you'll see this type of format change and know your information will most likely be securely encrypted.

**Stealth diagnostics**    This a diagnostic process used to detect attacks on a computer device or network.

**TCP/IP**    Transmission Control Protocol/Internet Protocol. This is the basic computer language that the IP uses to operate. TCP/IP is a two-layer program. The higher layer, Transmission Control Protocol, manages the assembling of a message or file into smaller packets that are transmitted over the Internet and received by a TCP layer that reassembles the packets into the original message. The lower layer, Internet Protocol, handles the address part of each packet so that it gets to the right destination. Each gateway computer on the network checks this address to see where to forward the message.

**Threat vectors**    This is any potential pathway that could initiate a cyber-attack. It could involve a fake or malicious website, the hijacking of an electronic session with colleagues, unsecured wireless local access networks (LAN) or wide area networks (WAN), e-mail links, unsecured mobile devices with antivirus security, social networks, malware on any electronic media, memory sticks, and any device that can be connected to your network such as via a USB connection.

**Trojan horse (or simply Trojan)**    A Trojan horse is a form of a computer virus that serves to create a secret or backdoor access to a user's device. (This is a hidden illicit and non-detected entry to a computer.) The hacker, once a Trojan horse is installed, can then have unauthorized access to the affected computer typically to steal data or passwords over a period of time and thus is less likely to be detected, since problems will likely occur over time rather than all at once. Trojans horses—or hidden back-doors—are not easily detectable. Computers, however, may appear to run slower. Malicious programs are classified as Trojans horses (or simply Trojans) if they do not attempt to inject themselves into other files (computer virus) or otherwise propagate themselves, which would then be labeled a "worm." A computer may host a Trojan via a malicious program. This is usually done by tricking a computer user into executing a installation command. This is often accomplished by opening an e-mail attachment disguised to not be suspicious. This might be in the form of a survey or access to a coupon or some other download. It might even be disguised as an antivirus program. A Trojan horse that is instructed to be activated at a particular time is sometimes called a "time bomb."

**Troll**   Someone that deliberately posts derogatory or inflammatory comments to a community forum, chat room, newsgroup, and/or a blog in order to bait other users into responding. It is also someone that frequents and eavesdrops on a chat room but does not contribute to it.

**Virus**   A computer virus is loaded without the knowledge of the computer or smart phone user. This malware can display unwanted messages or spam or do something much worse. This might be to corrupt or delete data on your computer, use your e-mail program to spread itself to other computers, or even erase everything on your hard disk. It might be in the form of a time bomb or Trojan horse and thus only reveal itself after weeks or even months have elapsed. Computer viruses are often spread by attachments in e-mail messages or instant messaging messages.

**VPN**   Virtual private network that is created to allow privacy of transmission over public networks.

**Website forgery**   This is another more sophisticated type of phishing technique. Website forgery is a malicious attack that creates the illusion of an exact replica of a trusted website on a victim's computer for the purpose of stealing personal information. By taking advantage of vulnerabilities in a victim's computer, a hacker can redirect the user to a forged site that appears completely trustworthy, including the browser displaying the correct URL with a secure connection, but in reality the website is a fake, with a deceptive graphical overlay masking the true URL.

**WEP encryption**   Wired equivalent privacy (WEP) is a security algorithm for IEEE 802.11 standard for wireless networks. WEP encryption was initially introduced as part of the original 802.11 standard ratified in September 1999; its intention was to provide data confidentiality comparable to that of a traditional wired network. WEP, recognized by the key of 10 or 26 hexadecimal digits, was at one time widely in use and was often the first security choice presented to users by router configuration tool. In 2003 the Wi-Fi Alliance formally voted to approve Wi-Fi Protected Access (WPA) as the new standard that should be used to replace WEP. In 2004, the new IEEE standard became 802.11i, which is known as WPA2.

**Whale phishing**   Another version of spear phishing is that known as whaling. This refers to a phishing attack against a senior-level individual within a commercial organization, institution, or governmental agency. Whalers will target a specific individual with access to sensitive systems and information. A typical attack would come in the form of an e-mail that appears to be from a customer, business partner, or even a more senior official. The message would likely refer to an urgent business matter in order to trick the victim into clicking on a malicious link or attachment or to reveal key passwords or data. If you work in a high level or key position within your company, you are at an elevated risk of being targeted. Even if you do not consider yourself to be in a senior role, you may still have access to sensitive information within your organization and be targeted as a result. As an agent of your employer, it is important to identify fraudulent e-mails and take every possible precaution against such attacks.

**Wi-Fi**    Wireless fidelity. This is a wireless local area network that is defined by the IEEE Standard 802.11. These wireless networks can be publicly available without a password or they can be password protected. To provide a reasonable level of privacy protection these Wi-Fi networks should have a reasonably high level of encryption.

**Wi-Fi Protected Access**    Wi-Fi Protected Access (WPA) and Wi-Fi Protected Access II (WPA2) are two security protocols and security certification programs developed by the Wi-Fi Alliance to secure wireless computer networks. The Alliance defined these in response to serious weaknesses researchers had found in the previous system, WEP (Wired Equivalent Privacy). A flaw in a feature added to Wi-Fi, called Wi-Fi Protected Setup, allows WPA and WPA2 security to be bypassed and effectively broken in many situations. WPA and WPA2 security implemented without using the Wi-Fi Protected Setup feature are unaffected by the security vulnerability.

**Wi-Fi Protected Setup (WPS)**    The addition of Wi-Fi Protected Setup to Wi-Fi networks creates a privacy and security vulnerability to WPA and WPA2.

**WLAN**    Wireless local area network. This is a generic name for any wireless data network where a router or Wi Fi system creates a "hotspot" for users to access. The IEEE standard 802.11i is the key specification for a Wi-Fi network.

**Worm**    A computer worm is a standalone malware computer program. The unique aspect of a worm is that it can self-replicate for the purpose of then spreading to other computers. Often, it uses the Internet to spread itself from the infected computer. Unlike a computer virus, a computer worm does not need to attach itself to an existing program and thus can be implanted anywhere in a computer's memory.

**Zip bomb**    A zip bomb, also known as a zip of death or decompression bomb, is a malicious archive file designed to crash or render useless the program or system reading it. It is often employed to disable antivirus software, in order to create an opening for more traditional viruses.

Rather than hijacking the normal operation of the program, a zip bomb allows the program to work as intended, but the archive is carefully crafted so that unpacking it (e.g., by a virus scanner) requires inordinate amounts of time, disc space, or memory. Most modern antivirus programs can detect whether a file is a zip bomb, to avoid unpacking it.

**"Zombie" computers**   This is a term that is applied to members of large botnets that have been assembled either to launch denial-of-service attacks, distribute e-mail spam on a very large scale, or conduct click fraud.

# Appendix B
# Current U. S. Priorities
# on Cybersecurity

## Five Key Objectives of Cybersecurity in the United States

1. Protecting the country's critical cyber infrastructure—our most important information systems—from threats.
2. Improving our ability to identify and report cyber incidents so that we can respond in a timely manner.
3. Engaging with international partners to promote Internet freedom and build support for an open, interoperable, secure, and reliable cyberspace.
4. Securing federal networks by setting clear security targets and holding agencies accountable for meeting those targets.
5. Shaping a cyber-savvy workforce and moving beyond passwords in partnership with the private sector.

## Why Cyberspace Is Crucial

Cyberspace touches nearly every part of our daily lives. It's the broadband networks beneath us and the wireless signals around us, the local networks in our schools and hospitals and businesses, and the massive grids that power our nation. It's the classified military and intelligence networks that keep us safe, and the World Wide Web that has made us more interconnected than at any other time in human history. We must secure our cyberspace to ensure that we can continue to grow the nation's economy and protect our way of life.

© Springer International Publishing Switzerland 2015
J.N. Pelton, I.B. Singh, *Digital Defense*, DOI 10.1007/978-3-319-19953-5

# Principles of U. S. Cybersecurity

The current Administration is employing the following principles in its approach to strengthen cybersecurity:

- Whole-of-government approach
- Network defense first
- Protection of privacy and civil liberties
- Public/private collaboration
- International cooperation and engagement

# The Five-Point Protection Plan

On February 12, 2013, President Obama signed Executive Order 13636, "Improving Critical Infrastructure Cybersecurity." This Executive Order is currently the most up-to-date statement of how the U. S. government is seeking to provide cyber security.

### #1. Protect Critical Infrastructure

WORKING WITH INDUSTRY: The government must work collaboratively with critical infrastructure owners and operators to protect our nation's most sensitive infrastructure from cybersecurity threats. Specifically, we are working with industry to increase the sharing of actionable threat information and warnings between the private sector and the U. S. government and to spread industry-led cybersecurity standards and best practices to the most vulnerable critical infrastructure companies and assets.

FRAMEWORK GUIDE FOR CYBER SECURITY: In 2014 the Administration launched a follow-on Cybersecurity Framework, a guide developed collaboratively with the private sector for private industry to enhance their cybersecurity. (See Chapter 7 for a detailed explanation of this framework.)

### #2. Improve Incident Reporting and Response

DETECT AND CHARACTERIZE: We must enhance our ability to detect and characterize cyber incidents, share information about them, and respond in a timely manner. These efforts encompass network defense, law enforcement, and intelligence collection initiatives, so we can better understand our potential adversaries in cyberspace.

AWARENESS AND RESPONSE: Detecting a cyber-threat or incident—and quickly acting on that information—are critical prerequisites to effective incident response. As directed in E.O. 13636, the U. S. government has developed systems and procedures to increase the timeliness and quality of cyber threat information shared with at-risk private sector entities. We are placing great emphasis on unity of effort by agencies with a domestic response mission.

### #3. Engage Internationally

INTERNATIONAL PARTNERSHIPS: Because cyberspace crosses every international boundary, we must engage with our international partners. We will work to create incentives for, and build consensus around, an international environment where states recognize the value of an open, interoperable, secure, and reliable cyberspace. We will oppose efforts to restrict Internet freedoms, eliminate the multi-stakeholder approach to Internet governance, or impose political and bureaucratic layers unable to keep up with the speed of technological change. An open, transparent, secure, and stable cyberspace is critical to the success of the global economy.

### #4. Pursue the Policy Objectives Laid Out in the U. S. International Strategy for Cyberspace

This includes:

- Developing international norms of behavior in cyberspace.
- Promoting collaboration in cybercrime investigations (Mutual Legal Assistance Treaty modernization).
- International cybersecurity capacity building

### #5. Secure Federal Networks

IMPROVE SECURITY: We must improve the security of all federal networks by setting clear targets for agencies and then hold them accountable to achieve those targets. We are also deploying improved technology to enable more rapid discovery of and response to threats to federal data, systems, and networks.

The Cybersecurity Cross Agency Priority (CAP) goal represents the Administration's highest cybersecurity priorities for securing unclassified federal networks.

## Shape the Future Cyber Environment

THE FUTURE: We are also looking to the future. We are working to develop a cyber-savvy workforce and ultimately to make cyberspace inherently more secure. We will prioritize research, development, and technology transition and harness private sector innovation while ensuring our activities continue to respect the privacy, civil liberties, and rights of everyone.

INNOVATION: The federal government is partnering with the private sector and academia to encourage and support the innovation needed to make cyberspace inherently more secure.

Source: https://www.whitehouse.gov/issues/foreign-policy/cybersecurity

# Appendix C
# The U.S. Comprehensive National Cybersecurity Initiative (CNCI)

(*Note*: This initiative was launched by President George W. Bush in January 2008 in National Security Presidential Directive 54/Homeland Security Presidential Directive 23 (NSPD-54/HSPD-23), and after an extensive review carried out after President Obama assumed office in January 2009 it was decided that the CNCI should be continued, strengthened, and made a part of an expanded U. S. cybersecurity policy.)

Cybersecurity has been identified as one of the most serious economic and national security challenges we face as a nation, but one that we as a government or as a country are not adequately prepared to counter. Shortly after taking office, President Obama therefore ordered a thorough review of federal efforts to defend the U. S. information and communications infrastructure and the development of a comprehensive approach to securing America's digital infrastructure.

In May 2009, the President accepted the recommendations of the resulting Cyberspace Policy Review, including the selection of an Executive Branch Cybersecurity Coordinator who will have regular access to the President. The Executive Branch was also directed to work closely with all key players in U. S. cybersecurity, including state and local governments and the private sector, to ensure an organized and unified response to future cyber incidents; strengthen public/private partnerships to find technology solutions that ensure U. S. security and prosperity; invest in the cutting-edge research and development necessary for the innovation and discovery to meet the digital challenges of our time; and begin a campaign to promote cybersecurity awareness and digital literacy from our boardrooms to our classrooms and begin to build the digital workforce of the twenty-first century. Finally, the President directed that these activities be conducted in a

© Springer International Publishing Switzerland 2015
J.N. Pelton, I.B. Singh, *Digital Defense*, DOI 10.1007/978-3-319-19953-5

way that is consistent with ensuring the privacy rights and civil liberties guaranteed in the Constitution and cherished by all Americans.

The activities under way to implement the recommendations of the Cyberspace Policy Review build on the Comprehensive National Cybersecurity Initiative (CNCI) launched by President George W. Bush in National Security Presidential Directive 54/Homeland Security Presidential Directive 23 (NSPD-54/HSPD-23) in January 2008. President Obama determined that the CNCI and its associated activities should evolve to become key elements of a broader, updated national U. S. cybersecurity strategy. These CNCI initiatives will play a key role in supporting the achievement of many of the key recommendations of President Obama's Cyberspace Policy Review.

The CNCI consists of a number of mutually reinforcing initiatives with the following major goals designed to help secure the United States in cyberspace:

To establish a front line of defense against today's immediate threats by creating or enhancing shared situational awareness of network vulnerabilities, threats, and events within the federal government—and ultimately with state, local, and tribal governments and private sector partners—and the ability to act quickly to reduce our current vulnerabilities and prevent intrusions.

To defend against the full spectrum of threats by enhancing U. S. counterintelligence capabilities and increasing the security of the supply chain for key information technologies.

To strengthen the future cybersecurity environment by expanding cyber education; coordinating and redirecting research and development efforts across the federal government; and working to define and develop strategies to deter hostile or malicious activity in cyberspace.

In building the plans for the CNCI, it was quickly realized that these goals could not be achieved without also strengthening certain key strategic foundational capabilities within the government. Therefore, the CNCI includes funding within the federal law enforcement, intelligence, and defense communities to enhance such key functions as criminal investigation; intelligence collection, processing, and analysis; and information assurance critical to enabling national cybersecurity efforts.

The CNCI was developed with great care and attention to privacy and civil liberties' concerns in close consultation with privacy experts across the government. Protecting civil liberties and privacy rights remain fundamental objectives in the implementation of the CNCI.

In accord with President Obama's declared intent to make transparency a touchstone of his presidency, the Cyberspace Policy Review identified enhanced information sharing as a key component of effective cybersecurity. To improve public understanding of federal efforts, the Cybersecurity Coordinator has directed the release of the following summary description of the CNCI.

# CNCI Initiative Details

## Initiative #1. Manage the Federal Enterprise Network as a Single Network Enterprise with Trusted Internet Connections

The Trusted Internet Connections (TIC) initiative, headed by the Office of Management and Budget and the Department of Homeland Security, covers the consolidation of the federal government's external access points (including those to the Internet). This consolidation will result in a common security solution which includes: facilitating the reduction of external access points, establishing baseline security capabilities; and, validating agency adherence to those security capabilities. Agencies participate in the TIC initiative either as TIC Access Providers (a limited number of agencies that operate their own capabilities) or by contracting with commercial Managed Trusted IP Service (MTIPS) providers through the GSA-managed NETWORX contract vehicle.

## Initiative #2. Deploy an Intrusion Detection System of Sensors Across the Federal Enterprise

Intrusion detection systems using passive sensors form a vital part of U. S. government network defenses by identifying when unauthorized users attempt to gain access to those networks. DHS is deploying, as part of its EINSTEIN 2 activities, signature-based sensors capable of inspecting Internet traffic entering federal systems for unauthorized accesses and malicious content. The EINSTEIN 2 capability enables analysis of network flow information to identify potential malicious activity while conducting automatic full-packet inspection of traffic entering or exiting U. S. government networks for malicious activity using signature-based intrusion detection technology. Associated with this investment in technology is a parallel investment in manpower with the expertise required to accomplish DHS's expanded network security mission. EINSTEIN 2 is capable of alerting US-CERT in real-time to the presence of malicious or potentially harmful activity in federal network traffic and provides correlation and visualization of the derived data. Due to the capabilities within EINSTEIN 2, US-CERT analysts have a greatly improved understanding of the network environment and an increased ability to address the weaknesses and vulnerabilities in federal network security. As a result, US-CERT has greater situational awareness and can more effectively develop and more readily share security-relevant information with network defenders across the U. S. government, as well as with security professionals in the private sector and the American public. The Department of Homeland Security's Privacy Office has conducted and published a Privacy Impact Assessment for the EINSTEIN 2 program.

## Initiative #3. Pursue Deployment of Intrusion Prevention Systems Across the Federal Enterprise

This Initiative represents the next evolution of protection for civilian departments and agencies of the federal Executive Branch. This approach, called EINSTEIN 3, will draw on commercial technology and specialized government technology to conduct real-time full-packet inspection and threat-based decision-making on network traffic entering or leaving these Executive Branch networks. The goal of EINSTEIN 3 is to identify and characterize malicious network traffic to enhance cybersecurity analysis, situational awareness, and security response. It will have the ability to automatically detect and respond appropriately to cyber threats before harm is done, providing an intrusion prevention system supporting dynamic defense. EINSTEIN 3 will assist DHS US-CERT in defending, protecting, and reducing vulnerabilities on Federal Executive Branch networks and systems. The EINSTEIN 3 system will also support enhanced information sharing by US-CERT with federal departments and agencies by giving DHS the ability to automate alerting of detected network intrusion attempts and, when deemed necessary by DHS, to send alerts that do not contain the content of communications to the National Security Agency (NSA) so that DHS efforts may be supported by NSA exercising its lawfully authorized missions. This initiative makes substantial and long-term investments to increase national intelligence capabilities to discover critical information about foreign cyber threats and use this insight to inform EINSTEIN 3 systems in real time. DHS will be able to adapt threat signatures determined by NSA in the course of its foreign intelligence and DoD information assurance missions for use in the EINSTEIN 3 system in support of DHS's federal system security mission. Information sharing on cyber intrusions will be conducted in accordance with the laws and oversight for activities related to homeland security, intelligence, and defense in order to protect the privacy and rights of U. S. citizens.

DHS is currently conducting an exercise to pilot the EINSTEIN 3 capabilities described in this initiative based on technology developed by NSA and to solidify processes for managing and protecting information gleaned from observed cyber intrusions against civilian Executive Branch systems. Government civil liberties and privacy officials are working closely with DHS and US-CERT to build appropriate and necessary privacy protections into the design and operational deployment of EINSTEIN 3.

## Initiative #4. Coordinate and Redirect Research and Development (R&D) Efforts

No single individual or organization is aware of all of the cyber-related R&D activities being funded by the government. This initiative is developing strategies and structures for coordinating all cyber R&D sponsored or conducted by the U. S. government, both classified and unclassified, and to redirect that R&D where needed. This Initiative is critical to eliminate redundancies in federally funded cybersecurity research, and to identify research gaps, prioritize R&D efforts, and ensure the taxpayers are getting full value for their money as we shape our strategic investments.

## Initiative #5. Connect Current Cyber Ops Centers to Enhance Situational Awareness

There is a pressing need to ensure that government information security offices and strategic operations centers share data regarding malicious activities against federal systems, consistent with privacy protections for personally identifiable and other protected information and as legally appropriate, in order to have a better understanding of the entire threat to government systems and to take maximum advantage of each organization's unique capabilities to produce the best overall national cyber defense possible. This initiative provides the key means necessary to enable and support shared situational awareness and collaboration across six centers that are responsible for carrying out U. S. cyber activities. This effort focuses on key aspects necessary to enable practical mission bridging across the elements of U. S. cyber activities: foundational capabilities and investments such as upgraded infrastructure, increased bandwidth, and integrated operational capabilities; enhanced collaboration, including common technology, tools, and procedures; and enhanced shared situational awareness through shared analytic and collaborative technologies.

The National Cybersecurity Center (NCSC) within the Department of Homeland Security will play a key role in securing U. S. government networks and systems under this initiative by coordinating and integrating information from the six centers to provide cross-domain situational awareness, analyzing, and reporting on the state of U. S. networks and systems, and fostering interagency collaboration and coordination.

## Initiative #6. Develop and Implement a Government-Wide Cyber Counterintelligence (CI) Plan

A government-wide cyber counterintelligence plan is necessary to coordinate activities across all federal agencies to detect, deter, and mitigate the foreign-sponsored cyber intelligence threat to U. S. and private sector information systems. To accomplish these goals, the plan establishes and expands cyber CI education and awareness programs and workforce development to integrate CI into all cyber operations and analysis, increase employee awareness of the cyber CI threat, and increase counterintelligence collaboration across the government. The Cyber CI Plan is aligned with the National Counterintelligence Strategy of the United States of America (2007) and supports the other programmatic elements of the CNCI.

## Initiative #7. Increase the Security of Our Classified Networks

Classified networks house the federal government's most sensitive information and enable crucial war-fighting, diplomatic, counterterrorism, law enforcement, intelligence, and homeland security operations. Successful penetration or disruption of these networks could cause exceptionally grave damage to our national security. We need to exercise due diligence in ensuring the integrity of these networks and the data they contain.

## Initiative #8. Expand Cyber Education

Although billions of dollars are being spent on new technologies to secure the U. S. government in cyberspace, it is the people with the right knowledge, skills, and abilities to implement those technologies who will determine success. However there are not enough cybersecurity experts within the federal government or private sector to implement the CNCI, nor is there an adequately established federal cybersecurity career field. Existing cybersecurity training and personnel development programs, while good, are limited in focus and lack unity of effort. In order to effectively ensure our continued technical advantage and future cybersecurity, we must develop a technologically skilled and cyber-savvy workforce and an effective pipeline of future employees. It will take a national strategy, similar to the effort to upgrade science and mathematics education in the 1950s, to meet this challenge.

## Initiative #9. Define and develop enduring "leap-ahead" technology, strategies, and programs

One goal of the CNCI is to develop technologies that provide increases in cybersecurity by orders of magnitude above current systems and which can be deployed within 5–10 years. This initiative seeks to develop strategies and programs to enhance the component of the government R&D portfolio that pursues high-risk/high-payoff solutions to critical cybersecurity problems. The federal government has begun to outline Grand Challenges for the research community to help solve these difficult problems that require 'out of the box' thinking. In dealing with the private sector, the government is identifying and communicating common needs that should drive mutual investment in key research areas.

## Initiative #10. Define and Develop Enduring Deterrence Strategies and Programs

Our nation's senior policymakers must think through the long-range strategic options available to the United States in a world that depends on assuring the use of cyberspace. To date, the U. S. government has been implementing traditional approaches to the cybersecurity problem—and these measures have not achieved the level of security needed. This Initiative is aimed at building an approach to cyber defense strategy that deters interference and attack in cyberspace by improving warning capabilities, articulating roles for private sector and international partners, and developing appropriate responses for both state and non-state actors.

## Initiative #11. Develop a Multi-pronged Approach for Global Supply Chain Risk Management

Globalization of the commercial information and communications technology marketplace provides increased opportunities for those intent on harming the United States by penetrating the supply chain to gain unauthorized access to data, alter data, or interrupt communications. Risks stemming from both the domestic and globalized supply chain must be managed in a strategic and comprehensive way over the entire lifecycle of products, systems, and services. Managing this risk will require a greater awareness of the threats, vulnerabilities, and consequences associated with acquisition decisions; the development and employment of tools and resources to technically and operationally mitigate risk across the lifecycle of products (from design through retirement); the development of new acquisition policies and practices that reflect the complex global marketplace; and partnership with industry to develop and adopt supply chain and risk management standards and best practices. This initiative will enhance federal government skills, policies, and processes to provide departments and agencies with a robust toolset to better manage and mitigate supply chain risk at levels commensurate with the criticality of, and risks to, their systems and networks.

## Initiative #12. Define the Federal Role for Extending Cybersecurity into Critical Infrastructure Domains

The U. S. government depends on a variety of privately owned and operated critical infrastructures to carry out the public's business. In turn, these critical infrastructures rely on the efficient operation of information systems and networks that are vulnerable to malicious cyber threats. This Initiative builds on the existing and ongoing partnership between the federal government and the public and private sector owners and operators of Critical Infrastructure and Key Resources (CIKR). The Department of Homeland Security and its private-sector partners have developed a plan of shared action with an aggressive series of milestones and activities. It includes both short-term and long-term recommendations, specifically incorporating and leveraging previous accomplishments and activities that are already underway. It addresses security and information assurance efforts across the cyber infrastructure to increase resiliency and operational capabilities throughout the CIKR sectors. It includes a focus on public-private sharing of information regarding cyber threats and incidents in both government and CIKR.

https://www.whitehouse.gov/issues/foreign-policy/cybersecurity/national-initiative

# Appendix D
# Cybersecurity Activities and Policies Around the World

## The European Union Agency for Network and Information Security (ENISA)

### Mission Statement

The ENISA is charged with assisting European Union states and the European Commission to better understand the emerging Critical Information Infrastructure Protection (CIIP) landscape and to issue important recommendations to influence the policy process in areas such as smart grids, ICS-Supervisory Control and Data Acquisition (SCADA), interconnected networks, Cloud computing, botnets, and mutual aid agreements.

© Springer International Publishing Switzerland 2015
J.N. Pelton, I.B. Singh, *Digital Defense*, DOI 10.1007/978-3-319-19953-5

UNISA also helps to develop good practices in areas such as national contingency plans, cybersecurity strategies, minimum security measures for ISPs, national cyber exercises, trusted information sharing, and others.

UNISA organizes complex, multi-national, and multi-stakeholder cyber exercises (e.g., Cyber Europe 2010, Cyber Atlantic 2011, Cyber Europe 2012, and most recently Cyber Europe 2014). UNISA also offers training and seminars to EU states in areas of its competence, such as national exercises, contingency plans, and incident reporting. Finally, UNISA assists National Telecom Regulatory Authorities in implementing a harmonized concept on mandatory incident reporting.

UNISA serves as a co-manager with the Commission the Pan European Public Private Partnership for Resilience (EP3R) to facilitate the dialogue among public and private stakeholders on emerging CIIP issues. It contributes to the European Commission's policy and strategic initiatives (e.g., Internet security strategy) and verifying that such recommendations are properly addressed by all concerned stakeholders.

The following questions and answers explain what functions UNISA does and does not perform.

## What Does ENISA Do?

ENISA's role is to enhance the cybersecurity prevention work and capability of the European Union and its member states, and as a consequence, the business community to prevent, address and respond to network and information security challenges. To this end agency activities are focused on:

- advising and assisting the commission and the member states on information security and in their dialog with industry to address security-related problems in hardware and software products;
- collecting and analyzing data on security incidents in Europe and emerging risks;
- promoting risk assessment and risk management methods to enhance our capability to deal with information security threats;
- awareness-raising and cooperation between different actors in the information security field, notably by developing public/private partnerships with industry in this field.

## What Does ENISA NOT Do?

ENISA's role is to act as a body of expertise in cybersecurity, NOT of being an inspecting, a directly operational, or regulating EU authority (in contrast to some other EU agencies). ENISA's remit clearly does not extend to the domains of operational national security, law enforcement, or defense, but remains in the prevention field. National and other EU bodies, e.g., EDPS and Europol, have the operational responsibilities for these matters. ENISA's reports and studies are usually used as starting point and input for the commission's initiatives and legislation in the field of cybersecurity.

## Why Was ENISA Created?

ENISA was created as it became increasingly clear to the member states that they were all making a lot of effort in this area. At the same time, the importance of making sure that the digital economy and information society was functioning properly became progressively more obvious. But in 2001, there was very little or no cooperation or information exchange between the member states, or between the governments and the industry in the field of information security. ENISA was set up to bridge this gap and bring forward good practices for all to use and to spread a culture of security across Europe.

By using the "open method" of coordination between the member states and the industry in this field, ENISA is facilitating and can contribute to a significant improvement in raising the exchange of information security knowledge and best practices between the member states. ENISA acts like a broker of knowledge and a switchboard of information. ENISA is also an EU point of contact for the external world on these matters, in close liason with the EEAS.

## What Does ENISA Do More Specifically?

The unit is responsible for assisting competent national EU agencies, the private sector, and the European Commission to develop sound and easily implemented strategies, policies, and measures for preparedness, response, and recovery that fully meet the emerging threats critical information infrastructures face today.

The unit fulfills its mission by the following:

- Assisting EU states and the commission to better understand the emerging CIIP landscape and issuing important recommendations to influence the policy process in areas such as smart grids, ICS-SCADA, interconnected networks, Cloud computing, botnets, and mutual aid agreements.
- Developing good practices in areas such as national contingency plans, cybersecurity strategies, minimum security measures for ISPs, national cyber exercises, trusted information sharing, and others.
- Organizing complex, multi-national and multi-stakeholder cyber exercises (e.g., Cyber Europe 2010, Cyber Atlantic 2011, Cyber Europe 2012, and Cyber Europe 2014).
- Offering training and seminars to EU states in areas of its competence, such as national exercises, contingency plans, and incident reporting.
- Assisting National Telecom Regulatory authorities in implementing a harmonized concept on mandatory incident reporting.
- Contributing to the commission's policy and strategic initiatives (e.g., Internet security strategy) and verifying that our recommendations are properly addressed by all concerned stakeholders.

## Who Is in Charge of ENISA?

ENISA is headed by the executive director, Dr. Helmbrecht (https://www.enisa.europa.eu/about-enisa/structure-organization/executive-director), who is responsible for all questions related to information security falling within the agency's remit. The work of the agency is overseen by a management board. The management board is composed of representatives from the EU member states, the European Commission as well as industry, academic, and consumer organization stakeholders. The executive director is moreover responsible to the European Parliament, the Council of the European Union, and the Court of Auditors. As ENISA's budget derives from the budget of the European Union, its expenditure remains subject to the normal EU financial checks and procedures.

## Why Is ENISA Situated in Crete?

As for the location of all the EU agencies (now 30 in number), this decision was taken by ministers from all EU countries. The objective is to locate an EU agency closer to EU's citizens in one of the member states. For ENISA, the ministers found a common agreement that ENISA should be situated in Greece. The Greek government then decided to situate ENISA in Crete, due to the close connection to one of the ten leading centers for Information and Communications Technology (ICT) in Europe, known as FORTH.

### How Does ENISA Communicate?

Communicating its results is key for ENISA to achieve impact. To do so, ENISA relies on the support of media and the EU member states as multipliers of information. Through its press releases and news items, ENISA publishes its key findings. Thereby, ENISA reaches out to all relevant actors and stakeholders in the member states, the EU institutions, the private sector and business, and other information security experts in the world, who subscribe to RSS feeds of PRs and news items.

Evidently, with a limited budget and staff, the ENISA website and social media tools are the main channels for acting like a 'switchboard' of information for the EU member states. The geographical location of ENISA, as for any EU agency, therefore, is of less relevance, as we have broadband connections in Crete and good support from the Greek authorities and all our stakeholders. We moreover reach out to the Information Security community through co-organizing conferences, and workshops.

### How Are the Industry's and Consumer's Opinions Taken into Account?

In its structure, include a permanent stakeholder's group and a management board that includes different stakeholders. Thereby, ENISA bridges the gap between the public and the private sectors in the field of information security.

### Is It Possible to Take Part in ENISA Studies/Do Business with ENISA?

As a European Union agency, our work and procurement of services and products, as well as in call for studies, is within strict, official procurement rules. All information concerning studies, or tenders launched through procurements by ENISA, is regularly updated under web announcements related to public procurement.

## How Many and Who Works at the Agency?

There are around 60 staff members working at ENISA. All are highly specialized and qualified from both the private and the public sector. All staff is recruited through EU-wide selections procedures, with applicants from across the 27 EU member states.

# Japanese Cybersecurity Initiatives

### Japanese Ministry of Defense Report of July 2013: Conclusions with Regard to Response to Cybersecurity Attacks

As no organization can singlehandedly defend itself from cyber-attacks, consider appropriate division of responsibilities among government ministries as well as strengthening coordination and cooperation with countries such as the United States and with the private sector. Additionally, consider policies to steadily introduce necessary equipment and train specialists. http://www.mod.go.jp/j/approach/agenda/guideline/2013_chukan/gaiyou_e.pdf

### Japanese Ministry of Defense (MoD) Cyber Defense Unit (CDU)

The Japanese Ministry of Defense (MoD) established a Cyber Defense Unit (CDU) on March 26, 2014, to detect and respond to attacks on the Ministry of Defense and the Japan Self-Defense Forces (JSDF). The CDU's objective is to help government and the JSDF to "deal effectively with the threat of cyber-attacks, which become more sophisticated and complex by the day." The CDU is tasked with monitoring Ministry of Defense and JSDF networks and will collaborate with other ministries and agencies in strengthening Japan's capability to respond to cyber threats. The unit will be located within MoD facilities and integrates about 90 JSDF personnel that previously undertook separate cyber-related activities in Japan's air, land and sea self-defense forces.

# OECD Guidelines: Towards a Culture of Security

## Preface

The use of information systems and networks and the entire information technology environment have changed dramatically since 1992, when the OECD first put forward the guidelines for the security of information systems. These continuing changes offer significant advantages to individual users who develop, own, provide, manage service and use information systems and networks ("participants"). Ever more powerful personal computers, converging technologies, and the widespread use of the Internet have replaced what were modest, stand-alone systems in predominantly closed networks. Today, participants are increasingly interconnected, and the connections cross national borders. In addition, the Internet supports critical infrastructures such as energy, transportation, and finance and plays a major part in how companies do business, how governments provide services to citizens and enterprises, and how individual citizens communicate and exchange information. The nature and type of technologies that constitute the communications and information infrastructure also have changed significantly. The number and nature of infrastructure access devices have multiplied to include fixed, wireless, and mobile devices, and a growing percentage of access is through "always on" connections. Consequently, the nature, volume, and sensitivity of information that is exchanged has expanded substantially. As a result of increasing interconnectivity, information systems and networks are now exposed to a growing number and a wider variety of threats and vulnerabilities. This raises new issues for security. For these reasons, these guidelines apply to all participants in the new information society and suggest the need for a greater awareness and understanding of security issues and the need to develop a "culture of security."

## Towards a Culture of Security

These guidelines respond to an ever-changing security environment by promoting the development of a culture of security—that is, a focus on security in the development of information systems and networks and the adoption of new ways of thinking and behaving when using and interacting within information systems and networks. The guidelines signal a clear break with a time when secure design and use of networks and systems were too often afterthoughts. Participants are becoming more dependent on information systems, networks, and related services, all of which need to be reliable and secure. Only an approach that takes due account of the interests of all participants, and the nature of the systems, networks, and related services can provide effective security.

Each participant is an important actor for ensuring security. Participants, as appropriate to their roles, should be aware of the relevant security risks and preventive measures, assume responsibility, and take steps to enhance the security of information systems and networks. Promotion of a culture of security will require both leadership and extensive participation and should result in a heightened priority for security planning and management, as well as an understanding of the need for security among all participants. Security issues should be topics of concern and responsibility at all levels of government and business and for all participants. These guidelines constitute a foundation for work towards a culture of security throughout society. This will enable participants to factor security into the design and use of all information systems and networks. They propose that all participants adopt and promote a culture of security as a way of thinking about, assessing, and acting on the operations of information systems and networks.

## *Aims*

These guidelines aim to:

- Promote a culture of security among all participants as a means of protecting information systems and networks.
- Raise awareness about the risk to information systems and networks; the policies, practices, measures and procedures available to address those risks; and the need for their adoption and implementation [9].
- Foster greater confidence among all participants in information systems and networks and the way in which they are provided and used.
- Create a general frame of reference that will help participants understand security issues and respect ethical values in the development and implementation of coherent policies, practices, measures, and procedures for the security of information systems and networks.
- Promote cooperation and information sharing, as appropriate, among all participants in the development and implementation of security policies, practices, measures, and procedures.
- Promote the consideration of security as an important objective among all participants involved in the development or implementation of standards.

# Principles

The following nine principles are complementary and should be read as a whole. They concern participants at all levels, including policy and operational levels. Under these guidelines, the responsibilities of participants vary according to their roles. All participants will be aided by awareness, education, information sharing and training that can lead to adoption of better security understanding and practices. Efforts to enhance the security of information systems and networks should be consistent with the values of a democratic society, particularly the need for an open and free flow of information and basic concerns for personal privacy. In addition to these security guidelines, the OECD has developed complementary recommendations concerning guidelines on other issues important to the world's information society. They relate to privacy (the 1980 OECD guidelines Governing the Protection of Privacy and Transborder Flows of Personal Data) and cryptography (the 1997 OECD guidelines for Cryptography Policy). These security guidelines should be read in conjunction with them.

#1. Awareness. Participants should be aware of the need for the security of information systems and networks and what they can do to enhance this security. Awareness of the risks and available safeguards is the first line of defense for the security of information systems and networks. Information systems and networks can be affected by both internal and external risks. Participants should understand that security failures may significantly harm systems and networks under their control. They should also be aware of the potential harm to others arising from interconnectivity and interdependency. Participants should be aware of the configuration of, and available updates for, their system, its place within networks, good practices that they can implement to enhance security, and the needs of other participants.

#2. Responsibility. All participants are responsible for the security of information systems and networks. Participants depend upon interconnected local and global information systems and networks and should understand their responsibility for the security of those information systems and networks. They should be accountable in a manner appropriate to their individual roles. Participants should review their own policies, practices, measures, and procedures regularly and assess whether these are appropriate to their environment. Those who develop, design and supply products and services should address system and network security and distribute appropriate information including updates in a timely manner so that users are better able to understand the security functionality of products and services and their responsibilities related to security.

#3. RESPONSE. Participants should act in a timely and cooperative manner to prevent, detect, and respond to security incidents in a timely manner. Recognizing the interconnectivity of information systems and networks and the potential for rapid and widespread damage, participants should share information about threats and vulnerabilities, as appropriate, and implement procedures for rapid and effective cooperation. Where permissible, this may involve cross-border information sharing and cooperation.

#4. ETHICS. Participants should respect the legitimate interests of others. Given the pervasiveness of information systems and networks in our societies, participants need to recognize that their action or inaction may harm others. Ethical conduct is therefore crucial, and participants should strive to develop and adopt best practices and to promote conduct that recognizes security needs and respects the legitimate interests of others.

#5. DEMOCRACY. The security of information systems and networks should be compatible with essential values of a democratic society. Security should be implemented in a manner consistent with the values recognized by democratic societies, including the freedom to exchange thoughts and ideas, the free flow of information, the confidentiality of information and communication, the appropriate protection of personal information, openness, and transparency.

#6. RISK ASSESSMENT. Participants should conduct risk assessments. Risk assessment identifies threats and vulnerabilities and should be sufficiently broad-based to encompass key internal and external factors, such as technology, physical and human factors, and policies and third-party services with security implications. Risk assessment will allow determination of the acceptable level of risk and assist in the selection of appropriate controls to manage the risk of potential harm to information systems and networks in light of the nature and importance of the information to be protected. Because of the growing interconnectivity of information systems, risk assessment should include consideration of the potential harm that may originate from others or be caused to others.

#7. SECURITY DESIGN AND IMPLEMENTATION. Participants should incorporate security as an essential element of information systems and networks. Systems, networks, and policies need to be properly designed, implemented, and coordinated to optimize security. A major, but not exclusive, focus of this effort is the design and adoption of appropriate safeguards and solutions to avoid or limit potential harm from identified threats and vulnerabilities. Both technical and non-technical safeguards and solutions are required and should be proportionate to the value of the information concerning the organization's systems and networks. Security should be a fundamental element of all products, services, systems, and networks, and an integral part of system design and architecture. For end users, security design and implementation consists largely of selecting and configuring products and services for their system.

#8. SECURITY MANAGEMENT. Participants should adopt a comprehensive approach to security management. Security management should be based on risk assessment and should be dynamic, encompassing all levels of participants' activities and all aspects of their operations. It should include forward-looking responses to emerging threats and address prevention, detection, and response to incidents, systems recovery, ongoing maintenance, review, and audit. Information system and network security policies, practices, measures, and procedures should be coordinated and integrated to create a coherent system of security. The requirements of security management depend upon the level of involvement, the role of the participant, the risk involved, and system requirements.

#9. REASSESSMENT. Participants should review and reassess the security of information systems and networks and make appropriate modifications to security policies, practices, measures, and procedures. New and changing threats and vulnerabilities are continuously discovered. Participants should continually review, reassess, and modify all aspects of security to deal with these evolving risks.

# About the Authors

**Dr. Joseph N. Pelton, Ph.D.** is a widely published award-winning author with some 40 books written, co-authored, or co-edited. His *Global Talk* won the Eugene Emme Literature Award and was nominated for a Pulitzer Prize. He is the co-author with Dr. Singh of the book *Future Cities,* published by the Intelligent Communities Forum in 2009, and *The Safe City: Living Free in a Dangerous World* in 2013.

Dr. Pelton is currently the principal of Pelton Consulting International. He is on the Executive Board of the International Association for the Advancement of Space Safety and chair of its International Academic Advisory Committee as well as the former President of the International Space Safety Foundation. He is the former Dean of the International Space University and Director Emeritus of the Space and Advanced Communications Research Institute (SACRI) at George Washington University.

Dr. Pelton was the Director of the Interdisciplinary Telecommunications Program at the University of Colorado from 1988 to 1997, and at the time it was the largest such graduate program in the United States. During his academic career Professor Pelton has taught at American University, the University of Colorado at Boulder, and George Washington University as well as serving as Dean at the International Space University. His undergraduate degree in physics is from the University of Tulsa and graduate degrees are from New York University and Georgetown University.

© Springer International Publishing Switzerland 2015
J.N. Pelton, I.B. Singh, *Digital Defense*, DOI 10.1007/978-3-319-19953-5

He previously held various executive positions at Intelsat and Comsat, including serving as Director of Project SHARE and Director of Strategic Policy for Intelsat. Intelsat's Project SHARE gave birth to the Chinese National TV University. Dr. Pelton was the founder of the Arthur C. Clarke Foundation and remains as the Vice Chairman on its Board of Directors.

He was also the founding President of the Society of Satellite Professionals (SSPI) and has been recognized in the SSPI Hall of Fame. He is on the board of the World Future Society and also frequently speaks and writes as a futurist. Dr. Pelton is a member of the International Academy of Astronautics, an Associate Fellow of the American Institute of Aeronautics and Astronautics (AIAA), and a Fellow of the International Association for the Advancement of Space Safety (IAASS).

He was the President of the Arlington County Civic Federation and was a member of its Long Range Planning Commission that initiated "smart growth" in Arlington. He is the immediate past Chair of the IT Advisory Commission for Arlington County that plays a key role in protecting the safety and resilience of the county's telecommunications and IT networks.

**Dr. Indu B. Singh, Ph.D.** is Vice President of Los Alamos Technical Associates (LATA) and head of its Washington, D.C. Operations. He manages U.S. federal, international, and commercial consulting and engineering services. Additionally, he leads the global cyber security business. Dr. Singh serves as Executive Director of LATA's Global Institute for Security and Training (GIST), which he founded in 2012. Dr. Singh was a Director at Deloitte Consulting LLP, and managed Deloitte's Systems Engineering and Weapons of Mass Destruction Practice. Previously, he served as a Managing Director at BearingPoint, Inc., a publicly traded company, which was later acquired by Deloitte Consulting LLP.

Dr. Singh is a pioneer in the designing and implementing of "smart cities" and "safe cities" around the world. He has led projects to design, build, and implement new cities and urban security and IT systems in Asia, the Middle East, and South America. Dr. Singh has led workshops in a number of areas, such as urban security systems, designing and building smart cities, and joined with Dr. Pelton at George Washington University in organizing a National Symposium on Security and Educational Needs for the Future.

In 2009 he joined with Dr. Pelton in writing *Future Cities* as a project for the Intelligent Community Forum headquartered in New York City. Dr. Singh also teamed with Dr. Pelton more recently in 2013 to write *The Safe City: Living Free in a Dangerous World*. Dr. Singh has published several other books on communications, IT systems, and security and was founding Editor-in-Chief of *Telematics and Informatics*, a global technology journal published by Elsevier B.V. He is a former faculty member of Rutgers University and has served as adjunct professor at American University and George Washington University. Dr. Singh resides in McLean, Virginia, USA.

# Technical Editor

**Alexander Pelton, J.D.** is an IT security and management consultant who is experienced advising federal cilents within the public sector in support of secure operational environments. Most recently he has served as Director at LongView international Technology solutions and is currently working as an independent consultant. His consulting background includes quality management, PMO support, organizational process improvement, IT strategy, governance, service-oriented architecture (SOA), risk management, systems integration using new and emerging tecnologies, and project management. He is a PMI certified project management professional (PMI), and holds a juris Doctor in Law, as well as a bachelor's in economics and business from the University of Colorado at Boulder.

# Index

© Springer International Publishing Switzerland 2015
J.N. Pelton, I.B. Singh, *Digital Defense*, DOI 10.1007/978-3-319-19953-5

sh p.133

Printed in the United States
By Bookmasters